SEVEN EXPERIMENTS THAT COULD CHANGE THE WORLD

"In the spirit of Charles Darwin, Sheldrake proposes seven experiments involving unexplained natural phenomena, based on common observations. . . . The author provides background, information-gathering methods, ways to expand the experiments further, and supporting material for reader participation."

Science News

"This is a hugely enjoyable book. It is also important one—as an exercise in the philosophy of schience and possibly as an insight into the material world."

The New Scientist

ALSO BY RUPERT SHELDRAKE

A New Science of Life

The Presence of the Past

The Rebirth of Nature

Dogs That Know When Their Owners Are Coming Home

Chaos, Creativity, and Cosmic Consciousness
(with Ralph Abraham and Terence McKenna)

Natural Grace
(with Matthew Fox)

The Physics of Angels
(with Matthew Fox)

SEVEN EXPERIMENTS THAT COULD CHANGE THE WORLD

A Do-It-Yourself Guide to Revolutionary Science

RUPERT SHELDRAKE

Park Street Press
Rochester, Vermont

For my children

Park Street Press
One Park Street
Rochester, Vermont 05767
www.InnerTraditions.com

Park Street Press is a division of Inner Traditions International

Library of Congress Cataloging-in-Publication Data

Sheldrake, Rupert.
Seven experiments that could change the world : a do-it-yourself
    guide to revolutionary science / Rupert Sheldrake.
p.   cm.
Originally published: New York: Riverhead, 1995.
Includes bibliographical references and index.
1. Science—Experiments—Popular works.   I. Title.
Q182.3.S48   1995   95–12047 CIP
507'.24—dc20

ISBN of the current edition: 0-89281-989-8

Printed and bound in the United States at Lake Book

10 9 8 7 6 5 4 3 2 1

This book was typeset in Bembo, with Schneidler as a display typeface

CONTENTS

PREFACE TO THE

SECOND EDITION

When this book was first published in 1994, it aroused a great deal of public interest, especially in relation to the tests with return-anticipating pets described in chapter 1. From readers of the book itself, and also through extensive publicity in the media, I received hundreds of letters from people about the perceptiveness of their animals. It soon became clear that many people had noticed behavior in dogs, cats, horses, parrots, and other domesticated animals that seemed to go beyond present scientific understanding. More than 3,500 people have now contributed case histories that are classified and stored on my computerized database.

Since 1994 I have been coordinating an extensive research program on the unexplained powers of animals, based on the first part of this book. Hundreds of videotaped experiments have shown that dogs are indeed able to anticipate their owners' return in a way that seems telepathic.

With the help of my research associates, I have interviewed dozens of people with special experience in animal behavior, including animal trainers, kennel and stable proprietors, zoo keepers, police dog handlers, and blind people with guide dogs. We have also carried out extensive surveys of hundreds of randomly selected households in Britain and the United States to find out how widespread various kinds of unexplained perceptiveness are in dogs, cats, and other domesticated animals. They turn out to be very common. I have summarized much of this research in my book *Dogs That Know When Their Owners Are Coming Home, and Other Unexplained Powers of Animals*, first published in 1999. In addition, my colleagues and I have published in scientific journals a variety of technical papers on this work, the details of which are given in the appendix to this edition.

Although the first chapter on domestic animals has received the greatest amount of publicity, most of the other lines of investigation I propose in the book have also been followed up by myself and others. Updates on progress in all these areas, together with references to publications in scientific journals, are given in the appendix.

The most popular of these areas of inquiry has been research on the sense of being stared at, along the lines described in chapter 4. Tens of thousands of such tests have been performed, many of them in schools and colleges. The results are overwhelmingly positive and hugely significant statistically.

In the first edition of this book I asked people to send their results to me by post, and I subsequently gave updates on this work through newsletters. These functions are now carried out through my Web site, www.sheldrake.org. I am very thankful to Matthew Clapp for his freely given services as Webmaster. The Web site also summarizes controversies and debates resulting from the research proposed in these chapters. Since the first publication of this book, I have in most cases improved on the experimental designs suggested. I have outlined these new methods in the appendix and intend to post further updates on my Web site.

I am grateful to the late Ben Webster of Toronto, the Institute of Noetic Sciences, the Lifebridge Foundation of New York, and the Bial Foundation of Portugal for their financial support.

PREFACE TO THE FIRST EDITION

I have been fascinated by some of the questions discussed in this book for many years, in the case of homing pigeons since early childhood. I have also spent more than twenty-five years engaged in scientific research, and have developed a great respect for the power of the experimental approach. I have seen for myself that through well-designed experiments one can ask questions of nature, and receive answers.

I have also been impressed by the way that fundamental research can be done on a shoestring budget. In the course of my scientific education at Cambridge I came across many examples of the "string and sealing-wax" tradition of British science. I encountered this tradition in a living form when for several years, as a Research Fellow of the Royal Society, I shared a laboratory in the Biochemistry Department at Cambridge University with the late Robin Hill, the doyen of research on photosynthesis, whose ongoing experiments cost even less than the standard budget allocated to first-year graduate students.

In India, where I spent more than five years doing agricultural research, I found that Indian scientists, through sheer necessity, have developed ingenious ways of conducting field research with minimum expense. At the international institute where I worked, near Hyderabad, I adapted and developed these local methods, mainly employing local villagers, and found this kind of research very productive. For example, I and my colleagues developed a new multiple-cropping system for pigeonpeas that is now widely used by farmers in India and contributes to an increased food supply.

More recently, my interest in the hypothesis of formative causation, first proposed in my book *A New Science of Life* (1981), has led me to apply the experimental method to asking unusual scientific questions, in particular about the buildup of habits in nature through the process of morphic resonance. Some of the early experiments to test this hypothesis are described in my book *The Presence of the Past* (1988); many more have been performed since at universities in Europe, America, and Australia. The results (which I plan to describe in a future book) are encouraging. I have been impressed by the elegant simplicity of experimental designs developed by the researchers, some of them students, who provide inspiring examples of doing far-reaching research at very low cost.

The idea of writing the present book arose in London in 1989. I had been invited to meet the Board of the Institute of Noetic Sciences (noetic means "pertaining to the mind"), a think tank based in California. They were planning a project on the nature of causation and invited me to give my views on the subject, especially in the light of the hypothesis of formative causation. As the discussion proceeded, we talked about future research programs in general. I was asked what I would do if I were a member of the Board and wanted to support interesting and productive research with limited resources. My answer was to draw up a list of simple, low-cost experiments that could change the world, and then to encourage this research program.

That evening, at dinner in the Garrick Club, several Board members, including a U.S. senator, suggested that I write a book on this very topic myself. This was a new idea for me, but the more I thought about it, the more I liked it. Outlines of new kinds of experiments kept coming to me, and, from the many I considered, I finally selected the seven in this book. So, this is not just a book, but a broad-based research program, with an open invitation to participate.

The development of these ideas has been assisted by a grant from the Institute of Noetic Sciences, of which I am a Senior Fellow, and this Institute has also offered to help coordinate this research program in North America. Additional financial support for the project has been provided by the Schweisfurth Foundation in Munich, Germany, thanks to the generosity of Mrs. Elizabeth Buttenberg.

I am indebted to many people for information, discussions, and advice about the various areas of research covered in this book, in particular Ralph Abraham, Sperry Andrews, Susan Blackmore, Jules Cashford, Christopher Clarke, Larry Dossey, Lindy Dufferin and Ava, Dorothy Emmet, Suitbert Ertel, Winston Franklin, Karl Geiger, Brian Goodwin, David Hart, Sandra Houghton, Nicholas Humphrey, Thomas Hurley, Francis Huxley, the late Brian Inglis, Rick Ingrasci, Stanley Krippner, Anthony Laude, David Lorimer, Terence McKenna, Dixie MacReynolds, Wim Nuboer, the late Brendan O'Regan, Brian Petley, Robbie Robson, Robert Rosenthal, Miriam Rothschild, Robert Schwartz, James Serpell, George Sirk, Dennis Stillings, Louis van Gasteren, Rex Weyler, and my wife, Jill Purce. I also received much valuable information from more than three hundred informants, experimenters, and correspondents, especially in connection with the behavior of pets, the homing of pigeons, the experience of phantom limbs, and the sense of being stared at. I am very grateful for all this freely given help.

I thank those who have read all or part of various drafts of this book, which has benefited from their criticism and comments, in particular Ralph Abraham, Christopher Clarke, Suitbert Ertel, Nicholas Humphrey, Francis Huxley, Brian Petley, Kit Scott, and my editors Christopher Potter and Andrew Coleman.

I am grateful to Christopher Sheldrake for doing the drawings in Figures 5, 7, and 8, and thank the following for permission to reproduce illustrations: Peter Bennett (Figure 1); Rick Osman (Figures 2 and 3); Jill Purce (Figure 4); Usborne Publishing Ltd (Figure 9B); and Stanley Krippner (Figure 12).

WHY BIG QUESTIONS

DON'T NEED

BIG SCIENCE

In this book, I propose seven experiments that could transform our view of reality. They would take us far beyond the current frontiers of research. They could reveal much more of the world than science has yet dared to conceive. Any one of them, if successful, would open up bewildering new vistas. Taken together, they could revolutionize our understanding of nature and ourselves.

This book is not only about a more open kind of science but about a more open way of *doing* science: more public, more participatory, less the monopoly of a scientific priesthood. The proposed experiments cost very little, and some practically nothing. This research is potentially open to anyone interested.

Because institutional science has become so conservative, so limited by the conventional paradigms, some of the most fundamental problems are either ignored, treated as taboo, or put at the bottom of the scientific agenda. They are anomalies; they don't fit in. For example, the direction-finding abilities of migratory and homing animals, such as monarch butterflies and homing pigeons, are very mysterious. They have not yet been explained in terms of orthodox science, and perhaps they cannot be. But direction-finding by animals is a low-status field of research, compared with, say, molecular biology, and very few scientists work on it. Nevertheless, relatively simple investigations of homing behaviour could transform our understanding of animal nature, and at the same time lead to the discovery of forces, fields, or influences at present unknown to physics. And such experiments need cost very little, as I show in this book. They are well within the capacity of many people who are not professional scientists. Indeed those best qualified to do this research would be pigeon fanciers, of whom there are more than five million worldwide.

In the past, most scientific research was carried out by amateurs; and amateurs, by definition, are people who do something because they love it. Charles Darwin, for example, never held any institutional post; he worked independently at his home in Kent, studying barnacles, writing, keeping pigeons, and doing experiments in the garden with his son Francis. But from the latter part of the nineteenth century onwards, science has been increasingly professionalized.[1] And since the 1950s, there has been a vast expansion of institutional research. There are now only a handful of independent scientists, the best known being James Lovelock, the leading proponent of the Gaia hypothesis, which is based on the idea that the Earth is a living organism. And although amateur naturalists and freelance inventors still exist, they have been marginalized.

Nevertheless, to explore areas which lie beyond the current boundaries of science has become much easier than most people imagine. Once again we are entering a phase of scientific development when pioneering investigations can be done by non-professionals, whether educated as scientists or not. Insofar as a scientific training is an advantage, there are millions of people all over the world who have had one. Computing power, once the monopoly of large organizations, is now widely available: there are computers in millions of homes. There are more people with leisure than ever before. Every year hundreds of thousands of students have to do scientific research projects as part of their training; some would welcome the chance to be real pioneers. And many informal networks and associations already provide models for self-organizing communities of researchers, working both within and outside scientific institutions. I envisage a complementary relationship between non-professional and professional researchers, the former having a greater freedom to pioneer new areas of research, and the latter a more rigorous approach, enabling new discoveries to be confirmed and incorporated into the growing body of science.

As in its most creative periods, science can once again be nourished from the grass roots up. Research can grow from a personal interest in the nature of nature—an interest which originally impels many people into scientific careers but is often smothered by the demands of institutional life. Fortunately, an interest in nature burns as strong, if not stronger, in many people who are not professional scientists.

Probably most readers of this book will not have the time or inclination actually to do the proposed experiments. But the very idea that they *could* participate is empowering, and I have found it is warmly welcomed both by those with a scientific education and those without. I have also found that by proposing particular experiments, the discussion of a topic is immediately sharpened, and the questions better focused.

Within the natural sciences, from time to time revolutions overturn established orthodoxies.[2] But at the heart of science is the experimental method. This remains central while paradigms come and go. Although I am convinced there is much wrong with the present state of science, I am a firm believer in the importance of experiments. Otherwise I would not be writing this book.

There is nothing particularly mysterious about the experimental method. It is a specialized form of a fundamental process found in all human societies, and indeed throughout the animal kingdom, namely learning by experience. The Latin word *experire,* to try out, is the root of our English words "experience" and "experiment" (and also "expert" and "expertise"). In French, *expérience* means both experience *and* experiment, as does the Greek *empeiros,* the source of our word "empirical."

Scientific experiments are deliberately and consciously contrived to give answers to questions. Experiments are ways of questioning nature. They can be used to decide between rival hypotheses, by allowing nature herself to speak through the data. Experiments are in this sense modern forms of oracles. The traditional diviners and interpreters of oracles included shamans, soothsayers, sages, seers, prophets and prophetesses, priests and priestesses, witches and magicians. In the modern world, scientists have taken on many of these roles.

Scientific hypotheses are tested through observation, and the best hypotheses are those that fit the observations best. Only through experiments can our understanding of nature be advanced; only through empirical evidence can a new scientific paradigm be established; only through experimental testing can science progress. This faith in the experimental method is fundamental to the practice of science and is shared by practically all scientists, myself included.

There has rarely been more public interest in the fundamental questions of science—for example in cosmology, quantum theory, chaos, complexity, evolution, consciousness—but at the same time there has never been more public alienation from official research. This book draws attention to areas of research neglected as a result of conventional habits of thought, where relatively simple experiments could yield rich returns, with extraordinary opportunities for breakthroughs worthy of the name. Inexpensive experiments open up pioneering research to non-professionals, and at the same time provide new opportunities for professional researchers faced with ever-increasing difficulties in finding funds, as well as students in search of exciting projects.

In Britain, research on the topics proposed in this book is being coordinated by the Scientific and Medical Network; in the United States by the Institute of Noetic Sciences (See U.S. address on p. 252) and coordinating centers have also been established in France, Germany, Holland, and Spain. These centers will help to put researchers in touch with each other, offer advice on experimental methods and statistical procedures, and provide regular updates through newsletters.

PART ONE

EXTRAORDINARY POWERS OF ORDINARY ANIMALS

WHY PUZZLING POWERS

OF ANIMALS HAVE

BEEN NEGLECTED

Institutional biology is currently dominated by the mechanistic theory of life, according to which all animals and plants are essentially complex machines, in principle fully explicable in terms of ordinary physics and chemistry. This theory is far from new. It was first proposed in the seventeenth century by René Descartes as part of the mechanistic philosophy of nature: the cosmos was a machine, and so was everything within it, including human bodies. Only the conscious, rational mind of man was different, being spiritual in essence. The mind was supposed to interact with the machinery of the body through a small region of the brain.

In many ways the mechanistic approach to life has been effective. Factory farming, agribusiness, genetic engineering, biotechnology, and modern medicine all bear tribute to its practical utility. And in terms of fundamental understanding, much has been learned about the molecular basis of living organisms, the nature of the genetic material, DNA, the chemical and electrical activities of the nervous system, the physiological role of hormones, and so on.

Academic biology has also inherited from seventeenth-century science a strong faith in reductionism: more complex systems should be explained in terms of smaller and simpler parts. Originally it was believed that the atoms formed the fundamental bedrock for all physical explanation. Now that atoms are known to be complex structures of activity composed of subatomic particles, themselves patterns of vibration within fields, the seemingly solid foundations of materialistic science have dissolved. In the words of the philosopher of science Karl Popper, "through modern physics materialism has transcended itself."[1] Nevertheless, in academic biology, the reductionist spirit remains strong and gives a great impetus to the attempt to reduce the phenomena of life to the molecular level. At this point, it is believed, the baton of reductionism can be passed to chemists, who in turn pass it on to physicists as molecules are reduced to atoms and finally to subatomic particles. Hence molecular biology is one of the most prestigious and well funded of the life sciences. Meanwhile, fields of inquiry that are inherently holistic have a low status in the hierarchy of science: for example, ethology, the study of animal behavior, or morphology, the study of the forms of organisms.

However, from the time that Descartes first proposed it, the mechanistic theory of life has been controversial, and until the 1920s it was opposed by a rival school of biology known as vitalism.[2] Vitalism is the doctrine that living organisms are truly alive. Mechanism is the doctrine that they are literally inanimate and soulless. For more than two centuries, vitalists argued that living organisms were animated by vital principles not known to physicists and chemists from the study of inanimate matter. By contrast, mechanists always claimed that there were no such things as vital factors or life forces. Their act of faith was that even if everything about living organisms could not yet be explained in terms of physics and chemistry, at some stage in the not-too-distant future it would be.

Because vitalists admitted the existence of unknown vital principles, they tended to be open-minded about the possibility of phenomena that could not be explained in mechanistic terms, such as psychic phenomena in humans and uncanny powers in animals.[3] By contrast, mechanists, as a matter of principle, were generally closed to the possibility of any phenomena that seemed inexplicable in terms of current physics and chemistry.

Mechanists often invoke an argument called "Occam's razor." This "razor" was originally used by a medieval Oxford philosopher, William of Occam, as a way of denying that theoretical constructs have any reality outside our minds. On the grounds that "entities are not to be multiplied unnecessarily," the simplest hypothesis is to be preferred. But when mechanists use Occam's razor, they do not do so in any strict philosophical sense, but merely as a justification for sticking to the currently orthodox scientific point of view.[4] They usually take it for granted that mechanistic explanations are the simplest, even though to attempt to apply them in practice to, say, predicting the behavior of an ant on the basis of the structure of its DNA would involve calculations so fiendishly complex that they could not be done. Any postulated non-material fields, forces, or principles are to be rejected—unless they have already been accepted by physicists. Mechanists have always feared, and still fear, that to admit the reality of anything "mysterious" or "mystical" in the realm of life would be to abandon the hard-won certainties of science.[5]

For those outside established science, these old controversies may seem dusty and remote. But unfortunately they are still relevant today. Most biologists, agriculturalists, and doctors have been brought up to believe that the mechanistic theory represents the triumph of reason over superstition, from which true science must be defended at all costs. Nevertheless, psychic phenomena have refused to go away. Animals continue to behave uncannily. Non-mechanistic forms of medicine flourish outside the orthodox institutions. Popular doubts about the practical applications of mechanistic principles in factory farming, forestry, agribusiness, and vivisection are growing rather than diminishing. The prospect of genetic engineering excites more fear than admiration. And the mechanistic theory of evolution by blind chance and natural selection has failed to win the hearts and minds of most people, despite the strenuous efforts of neo-Darwinian evangelists.

All these factors conspire to produce a defensive attitude in many biologists, and an unwillingness to explore the possibility that life might be stranger than anything dreamed of in old-style physics. This helps to explain why the puzzling phenomena I discuss in the following three chapters have received so little attention from professional researchers.

Although the old vitalist-mechanist controversy has done much to shape the attitudes of present-day biologists, it is no longer, in my opinion, a fruitful way to explore the nature of life. Since the 1920s, a broader alternative to the mechanistic theory of life has grown up in the form of the holistic or organismic philosophy of nature. From this point of view, the whole is more than the sum of its parts. Not only living organisms, but also non-biological systems, such as molecules, crystals, and galaxies, have holistic properties that are not reducible to their parts. Nature is made up of organisms, not machines.[6]

While academic biology is still under the sway of an old kind of thinking, a paradigm more than three centuries old, other branches of science have in many ways moved beyond the mechanistic worldview. Since the 1960s, the entire cosmos has looked more like a developing organism than a machine, continuously growing, and evolving new patterns of organization within itself as it does so. The rigid determinism of old-style physics has given way to a recognition of an inherent spontaneity in nature—through indeterminism at the quantum level, through non-equilibrium thermodynamics, and through the insights of chaos and complexity theories.[7] In cosmology, there has been the recognition of a kind of cosmic unconscious through the discovery of "dark matter," the nature of which is utterly obscure, but which nevertheless seems to constitute some 90–99 percent of the matter in the universe. Meanwhile, quantum theory has revealed strange and paradoxical aspects of nature, including the phenomenon of non-locality or non-separability, whereby systems that were once parts of a larger whole retain a mysterious connectedness even when many miles apart.[8]

Biologists in general take an old-fashioned view of physical reality. They have, by definition, specialized in biology; most have little or no education in quantum mechanics or other aspects of modern physics. Ironically, many are still hoping to reduce the phenomena of life to the physics of the past; but physics has moved on.

This ideological background helps to explain why the seemingly extraordinary powers of animals have been neglected by professional researchers, and consequently why such fundamental questions remain open. However, I am not advocating any particular theories to explain them. I believe the current orthodoxy is too limited, too narrow, but I also believe that the way forward depends on what nature herself tells us. At present we need more facts, and I hope that the following experiments will help to open up areas of inquiry that have been closed down for too long.

CHAPTER ONE

PETS WHO KNOW

WHEN THEIR OWNERS

ARE RETURNING

BONDS BETWEEN PETS AND PEOPLE

In my home town, Newark-on-Trent, one of my neighbors was a
widow who kept a cat. Her son was a merchant seaman. One day
she told me that she always knew when her son was coming home
on leave, even if he did not let her know when to expect him. The
cat went and sat on the front doormat and meowed for an hour or
two before he actually arrived, "so I always know when to start
getting his tea ready," she added.

She was not the kind of woman to make such things up, although this story may well have improved in the telling. Her mat-
ter-of-fact acceptance of this seemingly paranormal phenomenon
made me think. Was something strange really going on? Or was
this just some sort of illusion, a product of superstitious, wishful,
unscientific thinking? I soon found that many pet owners had similar tales to tell. In some cases the pets seemed to know hours
beforehand about the return of a long-absent member of the family. In other, more mundane examples, the pets would get excited
shortly before their owner came home from work.

In 1919, the American naturalist William Long published a fascinating book called *How Animals Talk,* in which he described how, when he was a schoolboy, he had a dog, an old setter named Don, who reacted to his time away at boarding school:

Don was left behind most unwillingly during my terms at school; but he always seemed to know when I was on my way home. For months at a stretch he would stay about the house, obeying my mother perfectly, though she never liked a dog; but on the day I was expected he would leave the premises, paying no heed to orders, and go to a commanding ledge behind the lane, where he could overlook the high-road. Whatever the hour of my coming, whether noon or midnight, there I would find him waiting. Once when I was homeward bound unexpectedly, having sent no word of my coming, my mother missed Don and called him in vain. Some hours later, when he did not return at his dinner time or answer her repeated call, she searched for him and found him camped expectantly in the lane. . . . And without a doubt that it would soon be needed, she went and made my room ready. If the dog had been accustomed to spending his loafing time in the lane, one might thoughtlessly account for his action by the accident or hit-and-miss theory; but he was never seen to wait there for any length of time except on the days when I was expected. And once he was observed to take up his watch within a few minutes of the hour when my train left the distant town. Apparently he knew when I headed homeward.[1]

Such stories abound. But how seriously can we take them? The sceptic within or without is always ready to explain them away in terms of chance coincidence, "subtle cues," sharp senses of smell and hearing, routine, or else the credulity, wishful thinking, and self-deception of those who want to believe in such things.

These standard armchair objections are not the result of detailed empirical studies. In fact, practically no research on the subject has yet been done. This lack of research is not because there is no general interest in the subject. On the contrary, there is a vast popular curiosity about the mysterious powers of pets. Neither has expense been the problem. The basic experiments cost practically nothing. Rather, scientific research has been prevented by a combination of three powerful taboos, which I consider in more detail at the end of this chapter: the taboo against investigating the paranormal; the taboo against taking pets seriously; and the taboo against experimenting with pets. For the time being I shall simply ignore these taboos and turn straightaway to a possible experiment.

Experiments with Animals Who Know When Their Owners Are Coming Home

The idea of a simple, inexpensive experiment to test how pets know when their owners are coming home came to me in a conversation with a sceptical friend, Nicholas Humphrey. I kept coming across stories about this intriguing phenomenon, and I asked him what he thought was going on. To my surprise he did not dispute the phenomenon itself; indeed he told me that his own dog seemed to have uncanny powers. But he was quick to add that there was nothing really mysterious going on; pets were good at responding to subtle cues and often had surprisingly sharp senses.

No doubt many people have had conversations like this. But rather than petering out inconclusively in the usual way, this particular one sparked off the idea for a simple experiment. If a pet responds well in advance of the arrival of its owner, the possibility that its behavior is explicable simply in terms of routine anticipation or sensory stimuli can be ruled out by coming home by an unusual means and at an unusual time. Moreover, to rule out the possibility that the pet is picking up the expectations of the person waiting at home, that person should not know when the absent member of the family is due to return.

This is not to say that routines, familiar sounds and smells, or the behavior of people at home are unimportant to the pet. I am sure they are very important. The purpose of the experiment is simply to tease apart the various influences that may normally work together, to find out whether there is an unexplained component of this behavior. Does the animal still know when the person is coming home when all conceivable sensory cues have been eliminated? In this sense, the experiment resembles research on pigeon homing. Even when cue after cue has been removed, the pigeons still get home (see chapter 2).

The only published research of this kind that I have come across was carried out by a "scientific friend" of William Long, the story of whose dog Don I have already quoted:

> This second dog, Watch by name and nature, was accustomed to meet his master much as Don met me in the lane. His owner, a busy carpenter and builder, had an office in town, and was accustomed to return from his office or work at all hours, sometimes early in the afternoon, and again long after dark. At whatever hour the man turned homeward, Watch seemed to follow his movement as if by sight; he would grow uneasy, would bark to be let out if he happened to be in the house, and would trot off to meet his master about halfway. . . . His strange "gift" was a matter of common knowledge in the neighborhood, and occasionally a doubtful man would stage an experiment: the master would agree to mark the hour when he turned homeward, and one or more interested persons would keep tabs on the dog. So my scientific friend repeatedly tested Watch, and observed him to take the road within a few moments of the time when his master left his office or building operations in the town, some three or four miles away.[2]

Of course, I would like to ask more questions about Watch and his behavior. But the dog and the people are dead. The only way forward is through observations and experiments with contemporary pets.

In 1992 I wrote an article on this subject, asking interested pet owners to contact me if they had any relevant observations, and especially if they would like to take part in research. This appeal was published in the "Members Research" section of the *Bulletin of the Institute of Noetic Sciences,* distributed to members of the Institute throughout North America and elsewhere.

I received more than a hundred letters in response, many full of fascinating information. Some observations already seemed to rule out an explanation in terms of routine reactions. Here, for example, is an account by Ms Louise Gavit of Morrow, Georgia:

In our case, there is no habit or schedule to my comings and goings, yet my husband tells me (and from past experience with two cats and one dog who did the same) our dog always responds to my coming home. In fact he seems to respond to my *intention and action* to come home. As near as I can measure my movements in comparison to the actions taken by the dog, his responses to my mental and physical actions are as follows: as I leave the place I have been and walk to my car with the intent to come home, our dog BJ will awaken from sleep, move to the door, lie down on the floor near the door, and point his nose toward the door. There he waits. As I near the drive, he becomes more alert and begins to pace and show excitement the nearer I move to home. He is always there to poke his nose through the crack, in greeting, as I open the door. This sensing seems to be unlimited by distance. He does not seem to respond at all to my leaving one place and moving to another, his response seems to be apparent at the time when I form the thought to return home, and take the action to walk toward my car to come home.

These are fascinating observations. I suggested to Ms Gavit that she try coming home by unusual means, for example by being driven by someone else in an unfamiliar car. She replied that this seemed to make no difference:

My method of travel is irregular: using my own car, my husband's car, a truck, or any number of cars driven by strangers to BJ, or walking. Somehow BJ responds to my thought/action just the same. Even when BJ has seen my car still inside the garage, which is located in the basement of my home, he responds.

Here is another example, from Mr. Starfire of Kahului, Hawaii:

My dog Debbie would always wait at the door about half an hour before my dad would get home from work. As my father was in the military he kept a very irregular work schedule. It did not matter if my father called ahead. For a while I suspected the dog responded to the phone. This was clearly not the case as sometimes my father would say he was coming home early but would have to stay late. Sometimes he wouldn't call at all. The dog was never fooled, so I ruled out the phone theory. My mother was the first to notice this behavior. She always started dinner when the dog went to the door. If the dog did not go to the door, then we knew our father would be coming home late. If he was late, the dog would still wait for him but not until he was on his way home.

Another example that cannot be explained in terms of routine expectations is from Mrs. Jan Woody of Dallas, Texas:

Our dog Cayce would know when my husband or myself was leaving to come home. She would stop what she was doing, whether in the yard (she would ask to come inside) or in the house, and go sit by the front door at the exact minute my husband or myself left whatever activity we were attending. Sometimes my husband would call me to say he was leaving the courthouse and ask if Cayce had gone to sit by the door. Other times, we'd tell each other when we had left and ask if she had gone to sit by the door at that minute. She figured that, like barking to let us know the mail had just been delivered, this was one of her tasks. The practice held true even if she was at my parents' home or at a motel or hotel. I don't see how she could hear our cars leave when they were in another city. I don't see how any sensory clues were provided to her, since neither my husband nor myself knew when the other was coming home (unless we called). Sometimes I'd stay a half hour or so later at an activity. Sometimes my husband's hearings would last all day, sometimes an hour.

Unfortunately, Cayce died in 1992, and so it is not possible to do any further tests to complement these remarkably clear observations by Mr. and Mrs. Woody.

Ms Vida Bayliss lives on forty acres of woodland in rural Oregon, three miles off the highway. Her dog Orion is a seven-year-old male, a Boxer-Dobermann mix, who ranges far and wide in the surrounding country. Yet when Ms Bayliss comes home, even though her schedule is "quite irregular," she finds he is nearly always there to greet her. I have heard many other stories of free-roaming dogs and cats who likewise seem to know when to be back home to greet their owners. Orion also distinguishes between family members and strangers before they arrive, barking to give warning of the approach of strangers but remaining silent with family members:

Orion also seems to be quite selective in who he decides is "part of the family." Since my divorce, even though my ex drives the same familiar vehicle, it is now cause for barking. But my parents, even though they visit infrequently, have always received the friendly, silent greeting. Once a family member drove up in a rented car and was barked at until a car window was opened and a greeting exchanged. Yet when my car is in the shop and I have a "loaner," I am not barked at. My driveway is rough, with three sharp switchbacks. Does Orion know me because I drive fast on this familiar road no matter what I drive?

To answer this question, Ms Bayliss could try coming home at an unusual time in an unfamiliar vehicle driven by someone else.

Dozens of other cases have been described to me by correspondents in the United States, and I have had more than thirty verbal reports of similar anticipatory behavior by dogs and cats in Britain and Germany. I have also been told of a parrot with similar powers. In each case, it has been possible to think of simple tests that could be done to find out more. The examples above illustrate the general principles.

Probably there are millions of people around the world who keep animals that seem to know when they are coming home. If only a few dozen were interested enough to do the basic pioneering research, it could soon be established whether or not this phenomenon goes beyond the conventional kinds of scientific explanation. If there is a seemingly "paranormal" effect, and if this is confirmed by a range of independent researchers, then further experiments could be done to investigate the phenomenon in more detail. At this stage the involvement of professional researchers would probably be helpful. And since sceptics would probably respond by thinking up ever-subtler alternative explanations, more sophisticated tests would need to be done to check any reasonable sceptical hypothesis. But a point might soon arrive when the sceptics' hypotheses become even more fantastic than the idea of a connection as yet unrecognized by science.

Research with pets who know when their owners are coming home is open to anyone who has such a pet, especially if they can count on the cooperation of family, friends and, of course, the animal itself. For students from homes with such pets, such research could be a science project of an extraordinarily interesting kind.

THE SOCIAL AND BIOLOGICAL BACKGROUND

In research on human parapsychology, most experiments soon become boring. Scores tend to fall off as the subjects begin to find the repetitive tests tedious. By contrast, the excited response of pets to the return of familiar people is repeated again and again; the animals are not bored by homecomings. For this reason I am optimistic about the possibility of obtaining repeatable and reliable results from experiments with pets who know when their owners are coming home.

Welcoming is a central feature of many person-pet relationships. In a survey in Cambridge, England, dog owners were asked to rate their animals in terms of twenty-two different aspects of canine behavior, such as playfulness, obedience, and affection. They were also asked to give their rating for a hypothetical "ideal" dog. Not surprisingly, this canine paragon loved going for walks, was obedient and intelligent, welcoming, expressive, and so on. But more interesting was the way in which the attributes of actual dogs matched their owners' ideal expectations:

> Prominent among these were the tendency to be very affectionate, to welcome the owner intensely whenever he or she came home, to be highly expressive (almost human), and to attend closely to everything the owner said or did. . . . Dogs and cats are particularly adept at conveying these apparent signals of friendship, and their ability to win friends and influence people owes more to their skill in this direction than any other factor. Perhaps the most obvious way in which these animals signal their liking for us is by their habit of seeking out our company and remaining near or even in

physical contact with us. The average dog, for example, behaves as if literally "attached" to its owner by an invisible cord. Given the opportunity, it will follow him everywhere, sit or lie down beside him, and exhibit clear signs of distress if the owner goes out and leaves it behind, or shuts it out of the room unexpectedly.[3]

Just as human greetings and reunions are usually conventional and ritualized, so are those of dogs. Many dogs produce breathless yelps of excitement; the sides of their mouth are drawn back in a so-called submissive grin; and unless they are well disciplined they try to jump up and lick their owner's face. In these and other ways they behave like pups greeting their parents, with their tail wagging so vigorously that the whole hindquarters become part of the movement. Wolf greetings are similar. When cubs are weaned, they are discouraged from suckling, and instead start soliciting food from their parents or other members of the pack. When the adult approaches with food in its mouth, they crowd excitedly around its head, wag their tails, adopt gestures of submission, and jump up and lick the corners of its mouth. In adult wolves, these same patterns of behavior develop into ritualized greetings and displays of pack solidarity. Most attention is directed to the highest-ranking animals, who act out the parental role by parading around with bones, sticks, or other objects in their mouths.[4]

Likewise, the greeting behavior of cats—often starting with a soft birdlike chirp, the cat approaching with tail up high, rubbing against the owner's hands or legs and purring loudly, and sometimes rolling on its back—is like that of kittens greeting their returning mother.

For many millions of years, among the wild ancestors of dogs and cats, the young remained behind while adults went off hunting. The same is true of their wild relatives today. The return of the hunters with food is an event of the most vital importance. Behind the greeting behavior of pets lies a long evolutionary history.

The close bonds between people and dogs date back over

10,000 years. Cats were more recently domesticated, starting perhaps 4,000 years ago in Egypt. If it turns out that there is a "paranormal" connection between pets and their owners, then it would be very likely that similar connections exist between members of groups in related wild species, and indeed in countless other animal species. And no one knows the nature of the social bonds in animal (or human) societies. I return to this question in chapter 3.

THREE TABOOS AGAINST RESEARCH WITH PETS

Although practically no research has yet been done with pets who know when their owners are coming home, the simple experiments discussed above would enable pioneering studies to be made at practically no expense. Why was this not done years ago? Because of powerful taboos, usually operating unconsciously. I now discuss these taboos briefly, because anyone considering research with pets needs to be aware of them. However, none of these taboos has much power when brought to consciousness, and they need present no obstacle to the research suggested in this chapter.

Taboo is an anglicized version of a Tongan word and refers to that which is "too sacred or too evil to be touched, named or used," or, in short, that which is forbidden.[5] Here are the three main taboos that have inhibited research on the unexplained powers of pets.

1. The Taboo Against Investigating the Paranormal

To start with, there is the general prohibition against taking "psychic" or "paranormal" phenomena seriously. If they really happen, they throw into doubt the mechanistic worldview, which is still the orthodoxy of institutional science. Therefore they are usually ignored or denied, at least in public.

This taboo is actively upheld by Skeptics. I am not referring here to the normal, healthy scepticism that is a component of common sense, but to self-proclaimed Skeptics (spelled with a capital S and a k), who form organized groups and serve as intellectual vigilantes, ready and willing to challenge any public claims of the paranormal.[6] Committed Skeptics tend to equate the mechanistic worldview with reason itself and are passionate in its defense. They are scientific fundamentalists. Their fear is that if claims of the paranormal are allowed to gain a foothold, scientific civilization will be swamped by an upsurge of superstition and religion. Their favorite approach is to dismiss "paranormal" phenomena as nonsensical, and to treat belief in them as an aberration arising from ignorance or wishful thinking—or, among those who ought to know better, as symptomatic of a weakness of the intellect.

Among respectable educated people, interest in "the paranormal" is treated like a kind of intellectual pornography. It flourishes in private, and in the less-reputable branches of the media, but it is more or less excluded from the educational system, from scientific and medical institutions, and from serious public discourse.

Unfortunately, many committed Skeptics confuse the defense of science with the defence of a particular worldview. Research of the kind I am suggesting in this chapter, or indeed in this book, may well lead beyond the mechanistic paradigm, but it is scientific none the less. It could lead to a larger and more inclusive scientific understanding of the world. On the other hand, if it shows that seemingly paranormal phenomena can in fact be explained in terms of standard scientific principles, then Skeptics will at last have some evidence to support their beliefs.

There is no need to be afraid of Skeptics. If they are opposed to empirical inquiry because of doctrinaire preconceptions, then they forfeit all claim to scientific credibility. But if they really believe in open-minded experimental research, as they profess, they should be a help rather than a hindrance.

2. The Taboo Against Taking Pets Seriously

The status of pets themselves involves a powerful taboo, which is pervasive and largely unconscious. The essential feature of this taboo is the vaguely defined notion that there is something strange, perverse, or wasteful in human affection for animals.

This taboo has recently been explored by James Serpell, a re-

searcher on animal behavior at Cambridge University. As a graduate student in the 1970s, he became interested in doing research on
the relations between people and their pets. He found to his surprise that there had been few, if any, scientific studies, in spite of
the fact that more than half the households in Western Europe and
North America contain at least one pet animal, including birds and
fish. In the European Community, there are an estimated 26 million pet dogs and 23 million pet cats. In the United States, there
are roughly 48 million dogs and 27 million cats, and annual expen-
diture on petfood and veterinary care is around 10 billion dollars.[7]
And as Nicholas Humphrey has pointed out: "In the United States
there are nearly as many dogs and cats as televisions. The effects of
television have been minutely researched and documented, but the
effect of pets remains virtually unanalyzed."[8] Why this extraordinary blindness on the part of scientific researchers?

Serpell's discussion is fascinating. He shows that the taboo is
related to the great gulf between the attitudes we have to pets and
our attitudes to other domesticated animals. Many dogs, cats, and
horses are loved and cherished, and are even mourned when they
die. But pigs, chickens, calves, and other animals grown in factory
farms are treated in the most brutal and exploitative way, devoid of
all affection. They are units in a production line; their only purpose
is to produce the maximum amount of food at the minimum cost.
Factory farms epitomize the mechanistic spirit. So does the use of
laboratory animals: they are treated as expendable, interchangeable
units for use in dispassionate experiments.

To justify this treatment, the less-favored animals have to be
regarded as inferior, unworthy of any sentimental attachment. A
terrible conflict arises if the exploited animals are considered to
have any value in themselves. One way to avoid this conflict is to
keep the privileged and the exploited animals in separate categories
in our minds. One category consumes petfood; the other is pro-
cessed into it. But if emotions spill over from pets to other animals,
there is trouble. People become vegetarians, or even animal rights
activists. The easiest solution is to denigrate people's relationships
with pets.

Prejudices against close relationships with animals are nothing new. In England, for example, during the persecution of witches, the witches' relationship with their animal "familiars," particularly cats, was considered perverted and wicked. But in modern industrial societies, the gulf between pets and other domesticated animals has been accentuated, general affluence enabling unprecedented numbers of pets to be kept in well-fed luxury, with no economic utility, for purely "subjective" reasons; meanwhile, in the "objective," outer world, countless less-favored animals are raised as mechanistically as possible in factory farms and laboratories.

This analysis makes it clear why pets are unsuitable for conventional mechanistic experimentation. Institutional science lies on the "objective" side of this divide, and pets are entirely alien to the mechanistic spirit. They are not expendable units but have individual personalities and form affectionate long-term relationships with people. They are hard to regiment. And they are not used to being treated "objectively" by detached experimenters who try to show no feelings, and neither are their owners. They inhabit the "subjective" world of private life, as opposed to the "objective" world of science.

Popular books on pets take for granted the importance of human-animal bonds. Here, for example, is some confident advice to pet owners by Barbara Woodhouse, author of several successful books on animal training.

> I believe one has to give a great deal of oneself to animals if one is to get the best out of them. And, what is more, one has to treat them as one would like to be treated oneself. If we are to get the best out of our dogs, it is no use shutting them up in kennels for the greater part of their lives and then expecting them to be intelligent when they come out. In my opinion, animals must live with one constantly, and learn words and thoughts that one says and transmits, if they are to be true companions.[9]

Meanwhile, in the United States, it is possible to go to workshops with your pet, to work on your relationship. There are pet counsellors, therapists, and healers, some even offering consultations by telephone. At the leading edge is Penelope Smith of Marin County, California, who gives workshops on how to increase telepathic communication with pets in a step-by-step program. Her basic message is the same as that of Barbara Woodhouse:

> Animals do understand what you say or think to them, that is, if you have their attention and they want to listen (just like anyone else). . . . The interesting thing is that the more you respect animals' intelligence, talk to them conversationally, include them in your life, and regard them as friends, the more intelligent and warm responses you'll usually get.[10]

In this context, there is little possibility of experimenting *on* pets, but there are many opportunities for a kind of research partnership *with* pets, not denying the emotional relationships between animals and people, but making this relationship the essence of the research itself.

3. The Taboo Against Experimenting with Pets

The third taboo is related to the second. Most pet owners are strongly attached to their animals and seek to protect them from harm. Science is seen as a generally negative force in relation to animals, particularly in drug testing and in vivisection. Every year millions of animals are sacrificed on the altar of science, including rabbits, guinea pigs, dogs, cats, and monkeys. (Oddly enough, "sacrifice" is actually used as a technical term in the scientific literature for the killing of an animal.) Science also has a negative image among many animal lovers for engendering the factory farming industry.

In this context, the idea that science could penetrate the sanctity of the home, subjecting beloved pets to its profane manipulations is deeply offputting. Interfering with pets is taboo.

Understandable though this reaction may be, it is not appropriate for the experiments I am suggesting here. These experiments do not involve cruelty or suffering. They should be fun not only for the people but for the animals too. And far from diminishing pets and their relationships with people, they are likely to lead to an increased respect for animals and their powers. Indeed, I think that one of the ways this particular kind of research could change the world is through giving a new sense of the living connections, seen and unseen, between the human and the animal realms.

FURTHER RESEARCH IN PARTNERSHIP WITH PETS

Knowing when their people are coming home is only one of the ways in which pets show surprising powers. There are several others, some of which provide yet more opportunities for fascinating low-cost research.

1. Homing abilities, as discussed in the next chapter.

2. Finding owners who have gone away from home, also discussed in the next chapter.

3. Apparent telepathic communication. In dramatic cases, some pets seem to know when their distant owner is in danger, reacting with signs of alarm and distress.[11] Other cases are more mundane: for example, some dogs seem to anticipate with uncanny accuracy when they are going to be taken for a walk. Some pets seem to know when their family is going to go away on holiday, even before they start packing. There are many stories about telepathic horses, and even some about telepathic tortoises. I recently received the following account from Ms Sharon Ronsse of Snohomish, Washington State:

> We have not been able to ascertain whether or not the tortoise knows (or cares) about our comings and goings. However, I have definitely noticed that the tortoise is telepathic when it comes to feeding him. I have concluded that his

behavior is not related to a habitual feeding schedule. I often feed him at odd times of the day and evening. When I first noticed that he would come out to the feeding area whenever I *thought* of feeding him, I started to do my own experiments on him. I found that at any time he was safely tucked in his little shelter and seemed to be snoozing, all I had to do was think about bringing out food. By the time I go to the kitchen and get something for him, he is out in the feeding area waiting for it.

Obviously pets are sensitive to subtle cues from people around them and can pick up influences of which their owners are unconscious. Experiments on apparent telepathic communication would need to eliminate these ordinary channels of communication, such as sight, hearing, and smell. For example, in the case of Ms Ronsse and her tortoise, the animal could be watched by someone else (or even monitored by means of a video camera). Meanwhile, inside the house, in a room in which she cannot be heard or seen by the tortoise, she thinks of feeding him (and then actually does so), according to a randomized schedule. Does the snoozing tortoise wake before she has started getting his food ready and before she has made any noise or movement?

4. Premonitions by animals of disasters. There are many stories about pets who try to prevent their owners going on journeys that turn out to be fatal. Even more dramatic is the behavior of animals before earthquakes. For example:

Before the Agadir earthquake in Morocco in 1960, stray animals, including dogs, were seen streaming from the port before the shock that killed 15,000 people. A similar phenomenon was observed three years later, before the earthquake which reduced the city of Skopje, Yugoslavia, to rubble. Most animals seemed to have left before the quake. The Russians observed, too, that animals began to abandon Tashkent before the 1966 earthquake.[12]

Clearly, an investigation of cases such as these could be of great practical value, and indeed in China such behavior by animals has been used successfully for centuries as an indicator of forthcoming calamities. However, this is obviously not an area where simple, harmless experiments would be easy.

5. Some pets returning from journeys seem to know when they are getting near home, even after a long ride in a car after dark when asleep. My wife and I had a cat, Remedy, who woke up when we were within a mile or two of home after sleeping contentedly for hours. Such a phenomenon could point to a direct connection between the animal and its home, perhaps related to the homing abilities discussed in the next chapter. Or it could simply indicate a response to a well-known pattern of movements and smells as the car approaches home by a familiar route. Or it could be a response to the changing behavior of the people in the car as they get ready to arrive.

Here again simple experiments could be very revealing. The hypothesis that the pet is responding to familiar stimuli can be tested by returning home by an unusual route, preferably one the pet has not traversed before. The possible influence of ambient sights, sounds, and smells can be minimized by keeping the pet in a box or basket, traveling after dark, keeping the windows closed, running the air conditioner, and playing music in the car. If under these conditions the pet shows no response, this would support an explanation in terms of familiar stimuli.

On the other hand, if the pets brought home by unusual routes still seem to know when they are nearing home, the next possibility to try to eliminate is an influence from the other people in the car. One way to do this would be to transport the pet in a van, under conditions where it could not see, hear, or smell its owner in the front. Its movements could either be recorded by an observer who did not know the destination of the van, or monitored by video, tape recorder, or other automatic means. Even better, the van could be driven by someone who did not know where the pet's home was, and who could not therefore emit subtle cues. The driver would simply be asked to follow a particular route which led past the pet's home but would not know in which street the home was.

If the pet still seemed to know when it was nearing home, then the hypothesis of a direct connection between the pet and its home would be strengthened. The nature of this connection and its possible relationship to homing behavior would be a subject for further research. But there would be no point in doing more sophisticated and expensive research until the existence of the phenomenon had been well established in the first place.

The purpose of this chapter is not to try to provide theories or explanations, but merely to show that the basic phenomena are still virtually uninvestigated. A scientific partnership with pets could lead to a great expansion of understanding, and a deeper appreciation of their powers of knowing.

CHAPTER TWO

HOW DO

PIGEONS

HOME?

A PERSONAL INTRODUCTION

When I was a young child, on Saturday mornings in the spring and summer my father took me to see a great liberation of pigeons. At the local railway station, racing birds from all over Britain were waiting in wicker baskets, arrayed in stacks. When the appointed time came, the porters opened the flaps, and out burst hundreds of pigeons, batch after batch, in a great commotion of wind and feathers (Figure 1). They flew up in the sky, circled around, and set off toward their faraway homes.

These birds were an inexhaustible source of fascination. As I got to know the porters, they let me help release the pigeons. Then when I was at primary school, I kept some homing pigeons of my own. But the pigeons were killed by a cat, and when I went away to boarding school I had no further opportunity for keeping birds.

Years later, in the early 1970s, when I was a Research Fellow at Clare College, Cambridge, my interest in pigeon homing reawoke, and I asked my colleagues in zoology how the pigeons did it. I soon found that no one really knew, an impression confirmed by reading the specialized papers and reviews in the scientific literature. Every reasonable hypothesis had been tried and seemed to have failed. I then saw that this intriguing mystery concerned not only homing, but also migration. How do English swallows migrate in the autumn to South Africa, and then in spring return to England, even to the very same building where they nested the year before? Again, no one knew.

I began to suspect that homing and migration might depend on a sense or power hitherto unrecognized by science. In particular, it seemed to me that there might be a direct connection between the birds and their home, rather like an invisible elastic band. I thought of a simple, inexpensive experiment to test this possibility, which I first tried out in Ireland in 1973. But I was unable to complete this research before I left for India in 1974 to take up a research post in an international agricultural institute. It was not until the 1980s, when I homed myself, that it was possible to begin working with pigeons again, this time in eastern England.

In this chapter, I first discuss what has so far been discovered about migration and homing in general, and about pigeons in particular. All explanations in terms of conventional senses and physical forces have by now been tested to destruction; our ignorance is more profound than ever. After summarizing the results of my own research, I conclude by outlining a potentially illuminating experiment, well within the capacity of many pigeon fanciers, pigeon clubs, and students at schools and colleges.

HOMING AND MIGRATION

Homing pigeons have been used for carrying messages for thousands of years. In the first book of the Bible, we read that a pigeon homed to Noah's ark with an olive leaf in its beak, showing Noah that the Flood was abating.[1] In ancient Egypt there was a carrier-pigeon postal system; in modern Egypt the pigeon is still the emblem of the Post Office. Even in the present century, pigeons have been widely used for carrying messages, not least by the military

FIGURE I

Racing pigeons being released from wicker carrying baskets at a railway station. (From an oil painting by Norman Fake, photographed by Peter Bennett.)

forces in the First and Second World Wars. Around the world today there are more than five million pigeon enthusiasts, who routinely race birds over distances of 500 miles or more. The sport is particularly popular in Belgium, Britain, Holland, Germany, and Poland. Pigeons can home from up to 700 miles in a single day, with average speeds exceeding 60 mph.

Pigeons are by no means alone in their homing abilities.[2] Countless anecdotes tell of domesticated animals, even cows, coming home after being left many miles away. The commonest concern dogs and cats. For example, a collie called Bobby, lost in Indiana, turned up at home in Oregon, more than 2,000 miles away, the following year.[3] Such cases form the basis for the well-known animal adventure story *The Incredible Journey*,[4] made into a film by Walt Disney, in which a Siamese cat, an old bull terrier, and a young Labrador find their way home through 250 miles of wild country in northern Ontario. The Labrador was the leader:

> It seemed as though he were never able to forget his ultimate purpose and goal—he was going home; home to his own master, home where he belonged, and nothing else mattered. This lodestone of longing, this certainty, drew him to lead his companions ever westward through wild and unknown country, as unerringly as a carrier pigeon.[5]

The human homing ability is most developed in nomadic peoples, where the sense of direction is essential to survival, as in the case of Australian Aborigines, the Bushmen of the Kalahari Desert in southern Africa, and the navigators of Polynesia.

The distance records are held by birds. Adélie penguins, Leach's petrels, Manx shearwaters, Laysian albatrosses, storks, terns, swallows, and starlings have all been known to home from more than a thousand miles away.[6] For example, when two Laysian albatrosses were taken from Midway Island in the central Pacific and released 3,200 miles away on the west coast of America, in Washington State, one returned in ten days, the other in twelve. A third came back from the Philippines, more than 4,000 miles away, in just over a month.[7] In an experiment with Manx shearwaters, birds were taken from their nesting burrows on the island of Skokholme, off the coast of Wales. One was released in Venice, Italy, and was back within fourteen days. Another returned in twelve and a half days from Boston, Massachusetts, a journey across the Atlantic of more than 3,000 miles.[8]

Clearly, these remarkable homing abilities are closely related to migrations from one home to another. In many cases, such as that of British swallows, migration is a double-homing system. They migrate in the autumn to their winter home in eastern parts of South Africa (where it is spring) and return to their British home in the northern spring.[9]

More amazing still is the instinctive ability of young birds to home to their ancestral winter quarters for the first time without the need to be guided by birds that have done it before. For instance, European cuckoos, raised by birds of other species, do not know their parents. In any case, the parent cuckoos leave in July or August for southern Africa, about a month before the new generation is ready to go. In due course the young cuckoos congregate and migrate in flocks to their African home to join their elders.

Even insects can migrate enormous distances to places they have never been before. The most famous is the monarch butterfly, which migrates between the United States and Mexico. In the autumn, when the previous generation has already died out, the new generation flies southward. Monarchs born near the Great Lakes in the eastern part of the United States, for instance, travel some 2,000 miles, overwintering in their millions on particular "butterfly trees" in the Mexican highlands. They die after breeding in their southern home. The next generation migrates northwards in the spring.[10]

How do migrating animals know where to go? In the case of migrant birds, the most popular hypothesis is that they orient by the stars, and perhaps are also exquisitely sensitive to the Earth's magnetic field. They are also supposed to have an inborn program, complete with star map and maybe also a magnetic map, that directs the migratory process. In the scientific literature, this is called an "inherited spatiotemporal vector-navigation program."[11] But little is really understood. This impressive-sounding technical term merely restates the problem, rather than solving it.

The main evidence for the role of stars is that when migratory birds are kept in cages in a planetarium at the beginning of the migration season, they tend to hop in the appropriate direction of migration according to the rotating pattern of "stars." But while stars may play a role, serving as a kind of compass, migrants can still find their way in the daytime or when the sky is heavily overcast.[12] For example, in a radar tracking experiment based in Albany County, New York, it was found that uninterrupted overcast skies lasting several days did not result in disorientation of nocturnal migrant birds of various species; there were "not even subtle changes in flight behavior."[13]

Fish can also migrate over hundreds or thousands of miles, and stars cannot explain their orientation. They must have other means of finding their way. Smell probably plays an important part when they are near their destination. In the case of salmon, there is good evidence that they "smell" their home river when they approach its estuary.[14] But smell cannot explain how they get near enough to the right stretch of coastline from thousands of miles away. Similar problems arise when trying to understand the migrations of marine turtles and other underwater migrants.

Both homing and migration are poorly understood, and an insight into one would illuminate the other. The investigation of migrations is difficult; it is far easier to work with homing behavior, especially in birds. Racing pigeons are the obvious choice. They have a strongly developed homing ability, and have been bred and selected for this ability over many generations. The techniques for keeping, breeding, and training them are well known, and they are relatively inexpensive.

Numerous experiments on homing have already been carried out with pigeons. Nevertheless, after nearly a century of dedicated but frustrating research, no one knows how pigeons home, and all attempts to explain their navigational ability in terms of known senses and physical forces have so far proved unsuccessful. Researchers in this field readily admit the problem. "The amazing flexibility of homing and migrating birds has been a puzzle for years. Remove cue after cue, and yet animals still retain some back-up strategy for establishing flight direction."[15] "The problem of navigation remains essentially unsolved."[16]

I now consider one by one the hypotheses proposed to explain pigeon homing, and show why they are all untenable.

DO PIGEONS REGISTER THE TWISTS AND TURNS OF THE OUTWARD JOURNEY?

How do pigeons taken hundreds of miles to an unfamiliar place know where their home is? How do they know which way to fly?

Charles Darwin was a keen pigeon fancier and kept a wide range of breeds.[17] In 1873, he tentatively proposed a hypothesis of pigeon homing in a paper in *Nature:* they do it by a kind of "dead reckoning," registering all the twists and turns of the outward journey, even when enclosed in a box.[18] In a subsequent paper in the same volume of *Nature,* J. J. Murphy proposed a mechanical analogy in terms of a ball hanging from the roof of a railway carriage, reacting to shocks given to it by changes in the carriage's direction and velocity:

> A machine could be constructed in connexion with a chronometer, for registering the magnitude and direction of all these shocks, with the time at which each occurred; and from these data, the position of the carriage, expressed in terms of distance and direction, might be calculated at any moment. . . . Further, it is possible to conceive of the apparatus as so integrating its results. . . . that they can be read off, without any calculation being needed.[19]

An updated technological analogy would be a computerized inertial navigation system. But in spite of these mechanical metaphors, it does not seem intrinsically plausible that racing pigeons, shut up in baskets, taken hundreds of miles inside trains, trucks, ships, or aeroplanes, subject to many twists and turns, can continuously compute their homeward direction with the highest precision.

In any case, this hypothesis has been tested and refuted. In 1893,

S. Exner showed that pigeons could still home perfectly well after being transported to the release site under heavy anaesthesia. More recent experiments with other species, such as herring gulls, have confirmed Exner's finding.[20] Nor does taking the pigeons to the point of release by complex and devious routes make them lose their way. They can even home after being transported in a large, light-proof rotating drum:

> The construction was unstable so that changes in the speed and direction of the transporting vehicle produced a momentary slowing of the drum. The outward journey through space was thus remarkably complicated by the irregularly varying rotation, about 1,200 rotations in the longest journey. Nevertheless, in every case the performance of the rotated birds, both in orientation and returns, was just as good as the untreated controls.[21]

In another series of experiments, carried out in Germany, on the outward journey the pigeons were rotated rather fast, at up to ninety revolutions per minute, in a variable magnetic field, unable to see out and isolated from any smells in the environments they were passing through. "Nevertheless, these pigeons were mostly as good at initial homeward orientation and homing performance as control birds were, which had been transported in open crates on top of the car."[22]

Finally, if the birds were to sense and integrate all the twists and turns of the journey, the appropriate organ would be the semicircular canals in the middle ear, which detect accelerations and rotations. The total destruction of this organ stops birds flying properly, but in experiments involving the surgical cutting of the horizontal canals, the pigeons still homed normally after being taken more than 200 miles away. Indeed they performed just as well as controls.[23] In other experiments, "pigeons with a variety of surgical lesions of the semicircular canals orient themselves accurately, whether they are tested under sunny conditions or overcast ones."[24] The inertial navigation hypothesis can therefore be ruled out, and is no longer seriously entertained by researchers in the field.[25]

Does Homing Depend on Landmarks?

It is sometimes suggested that homing depends on familiar landmarks. This is probably the case when the pigeons are released a few miles from the loft, and also when they home repeatedly over the same terrain. In one series of trials, when the birds were released for the fourth time at the same site, they appeared to be orienting by means of local landmarks. "By the seventh the knowledge of local landmarks was so good that the birds were able to, as it were, steeple-chase their way home. They acted as if they recognized that home was to be reached by flying towards landmark A, then toward landmark B, and so on."[26] We show a similar tendency ourselves. As we get to know new localities and routes, we too find our way around by familiar landmarks. But this is not how we find our way in the first place, *before* the landmarks are familiar.

In any case, pigeons can home from completely unfamiliar places, hundreds of miles from anywhere they have been before. After flying around in circles after being released, or even without circling, they generally set off in a homewards direction.[27] They can also home over the sea, even flying at night or in fog, as in the spectacular performances of some of the pigeons used by the Royal Air Force in the Second World War. Experienced racing birds, many of them volunteered by amateur pigeon enthusiasts, were carried on aircraft flying sorties to Germany over the North Sea. In the event of the aircraft being shot down, the crew, if they could, attached a message giving their location to the leg of one or more pigeons, released the birds, and hoped for the best.

The details of hundreds of extraordinary exploits are officially recorded in the Pigeon Roll of Honour, known as the Meritorious Performance List, and some birds were actually decorated for valiant service (Figure 2). Here is the official story of one medal-winning hen called White Vision, bred in Motherwell, Scotland, and based at RAF Sollum Voe in the Shetland Islands:

This pigeon was carried in a Catalina flying boat which, owing to engine failure, had to ditch in a rough sea in Northern waters at approximately 08.20 hours on the 11th

FIGURE II

"Winkie" and her awards. The account in the Meritorious Performance List is as follows: "On the 23rd February, 1942, damaged Beaufort ditched suddenly whilst returning from a strike off the Norwegian coast and partially broke up on heavy impact 120 miles from the Scottish coast. This pigeon escaped from the container accidentally in the crash and fell into the oily sea before struggling clear. Distance to base 129 miles, nearest land 120 miles, 1$^1/_2$ hours of daylight left. Pigeon homed soon after dawn next morning, exhausted, wet and oily. Air search for crew on very poor radio fix up to then unsuccessful. Sergeant Davidson, RAF Pigeon Service, deduced from the arrival of pigeon, its condition and other circumstances that the area of search was incorrect. Search was redirected in accordance with his advice and 15 minutes later the crew were located and rescue action taken. The rescued crew gave a dinner in honour of the pigeon and her trainer." (From Osman and Osman, 1976.)

October, 1943. Owing to radio failure, no SOS was received from the aircraft and no fix obtained, . . . At 17.00 hours "White Vision" arrived with a message giving the position and other information concerning the aircraft and crew. As a result sea search was continued in the direction indicated and at 00.05 hours the following morning the aircraft was sighted and the crew rescued. The aircraft had to be abandoned and sank. Weather conditions: visibility at place of release of pigeon 100 yards. Visibility at Base when pigeon arrived 300 yards. Head wind for pigeon 25 miles per hour. Heavy sea running, very low cloud, distance about 60 miles. Number of lives saved 11.[28]

The use of landmarks, and indeed any other visual cues, seems very unlikely to play a key role in homing feats such as this. Nevertheless, until the 1970s most attempts to explain homing in pigeons focused on vision as the key sense, if not for identifying landmarks then for navigating by the sun or even by the stars. All such visual hypotheses were ruled out in some remarkable experiments carried out in the United States at Duke University, North Carolina, and in Germany at Göttingen, in which pigeons were fitted with frosted-glass contact lenses. These so impaired their vision that they could not recognize familiar objects 20 feet away. Control birds were fitted with plain contact lenses.

When the birds with the frosted-glass lenses were released, "many refused to fly, hovered, or crash-landed nearby; others hit wires, trees or other obstacles. A certain proportion went high up into the sky and disappeared unusually high over the horizon." They flew in a peculiar way, with their bodies tilted upwards. This expression of "uncertainty" was recognized by hawks, which preyed with ease on such birds.[29] Some birds flew part of the way home and then perched, resting for longer or shorter periods.[30] But some found their way home from over 80 miles away. "Experimental birds usually arrived at the loft rather high in the sky, cautiously hovering down, a few hitting, most others missing, the loft. The birds could easily be caught by hand."[31] The birds had difficulty pinpointing the loft, suggesting that they needed their sense of vision for the final approach to the loft, which is hardly surprising. What is amazing is that they could get so close to home with their eyesight so severely impaired.

The leader of the Göttingen team, Klaus Schmidt-Koenig, summarized as follows the conclusions from a long series of experiments on pigeons with frosted-glass lenses, including the detailed tracking of homing birds by radio:

> For the navigational part of the homing flight, i.e., determining which direction is the home direction, visual cues turned out not to be essential. This navigational system is largely non-visual and guides the pigeon with amazing accuracy to the vicinity of the loft. The birds also seem to know when they are home and when they have missed the loft and the distance is again increasing.[32]

DO PIGEONS NAVIGATE BY THE SUN?

In the 1950s, the dominant hypothesis of pigeon homing was the "sun arc" theory of G. V. T. Matthews. He proposed that birds used a combination of the sun's elevation, together with its arc, extrapolated through the sky on the basis of observations of its movement, and also an accurate internal "chronometer." A pigeon taken south-west, for example, would find the sun's position unusually high and east (which is to say early), by an amount corresponding to its displacement from the loft. It could, in principle, "calculate" the home position.[33]

There are several strong arguments against this hypothesis. Pigeons can home under heavily overcast conditions; they can home with frosted-glass contact lenses on; and they can even home at night.[34] Moreover, they can still home when their time sense is severely disrupted, and Matthews's hypothesis requires a very accurate internal time-keeping process.

In a long series of experiments, pigeons had their internal

"clock" shifted by keeping them in darkness during the daytime and under artificial lights at night. For example, by turning on the lights six hours before dawn and plunging the pigeons into darkness six hours before sunset, within two weeks the pigeons had their internal "clock" set six hours early. When such birds were released, they set off about 90° to the left of the homewards direction. By contrast, birds whose "clock" had been set six hours late set off about 90° to the right of the homewards direction. Those with a twelve hour time shift set off in the opposite direction to home.[35]

At first sight such results seemed to confirm Matthews's theory. But in fact they show only that pigeons can use the sun's position as a kind of *compass*. And a compass is not sufficient to explain homing. Imagine being parachuted into a strange place with a watch but no map. From the position of the sun at different times of day, you could work out where north, south, east, and west were, but you would not know the direction of home.

In his hypothesis, Matthews claimed that pigeons used an internal clock, combined with the sun's position and the arc of its movement in the sky, not just as a compass but as a kind of *map*, enabling them to know the direction and distance of their home from the point of release. As well as failing to explain how birds can home at night and under heavily overcast skies, this hypothesis also failed to explain how time-shifted birds could still find their way home after the initial diversion caused by the false "sun-compass" reading.[36] And if time-shifted birds were released on overcast days, they were not confused but set off homewards and reached their loft as quickly as controls.[37]

Thus, on sunny days, the "sun-compass" of pigeons can play a part in their general sense of direction when they are released. But it cannot explain their ability to find their way home.

Does Homing Depend on Polarized Light
or Infrasound?

When the "sun arc" theory was in vogue, some people tried to explain the ability of pigeons to home on cloudy days in terms of a hypothetical response to the pattern of polarized light in the sky. Some insects, notably bees, are known to be sensitive to the polarization of light and can orient themselves if they can see patches of blue sky, even though the sun itself is obscured by cloud.

However, there are two fatal weaknesses in the polarized light hypothesis. First, even if pigeons could infer the sun's position from the pattern of polarization in blue patches, this would not explain their ability to home, because the sun's position and movement in the sky cannot explain homing, as we have just seen. Second, pigeons, unlike bees, are not sensitive to the polarization of light.[38]

Another unusual sensory ability sometimes invoked as a possible explanation for pigeon homing is infrasound. Pigeons are known from laboratory experiments to be unusually sensitive to low-frequency sounds. But this does not prove that they can hear their home from hundreds of miles away, or even from a few miles. The idea that they might home by means of infrasound is not even a hypothesis but just a vague and implausible suggestion. There is no evidence whatever to support it.

Does Homing Depend on Smell?

Mysterious abilities of animals are often explained, or explained away, in terms of a remarkable sense of smell. Pigeon homing is no exception, and over the last two hundred years smell has often been suggested as an explanation of homing. But a moment's reflection shows that this idea is implausible.[39] For example, consider the homing of racing pigeons from Spain to eastern England. Could birds released in Barcelona recognize where they were by sniffing the local smells, or by sniffing their home in Suffolk, England? And could they find the home direction by smell even when the wind was blowing toward rather than away from their home? Obviously not. The fact that birds can home to England from Spain flying with the wind, rather than against it, shows that the sense of smell cannot explain homing. This is particularly clear in northeastern Brazil, where trade winds blow from the south-east with little variation throughout the year. The local fanciers regularly and successfully race their birds from the south.[40]

An early version of the smell hypothesis proposed that pigeons have a special chemical sense organ in their air sacs. But then it was found that pigeons whose air sacs had been punctured by a needle still homed normally. The nasal cavities were next investigated, and birds had these cavities stuffed with wax. They could still home perfectly. All this was well established by 1915.[41]

The smell hypothesis, like the magnetic hypothesis, was revived in the 1970s when all else seemed to have failed. Floriano Papi and his colleagues in Italy proposed that pigeons build up an olfactory map of the surroundings of their home by associating odors with the direction of the wind. For example, if there is a pine forest to the north, they learn to associate pine forest smells with north winds. When transported to a release site, they have only to sniff the air to know the direction of home. In order to account for homing from great distances, where a local olfactory map would be of no assistance, Papi suggested that they registered the smells on the outward journey.

Papi's group built up a seemingly impressive body of evidence that their pigeons were indeed influenced by odors associated with wind direction.[42] For example, pigeons were raised with two different odors added to the wind, olive oil from the south and synthetic turpentine from the north. Then they were released with one of the odors applied to their nostrils, and were initially deflected from home as though the direction from which they had been released corresponded to the direction from which the odor was carried to them in the loft.[43]

Most attempts to repeat Papi's experiments in Germany and the United States gave very different results, with no detectable influence of odors.[44] However, even in Italy the sense of smell could not by itself have explained the pigeons' homing behavior. When they had been deliberately confused by the Italian scientists and set off in the wrong direction, they sooner or later corrected their course and returned home anyway. Indeed, many arrived almost as soon as control birds. Moreover, birds with plugged nostrils, severed olfactory nerves, or with tubes in their nostrils by-passing the olfactory epithelium could still fly home, although they tended to return slower than unmutilated control birds.

The Italians argued that the slower return of the mutilated birds confirmed the olfactory hypothesis.[45] Their sceptical colleagues in Germany and America suggested that it might simply be a general result of trauma. To test this idea, in Germany some of the pigeons had their olfactory epithelium anaesthetized with xylocaine, a potent local anaesthetic, blocking the sense of smell in a non-traumatic way. Sure enough, these pigeons flew off homewards when released, and returned as fast as controls.[46] In other experiments, anaesthesia with xylocaine reduced but did not prevent the ability of treated birds to home.[47]

The conclusion from this research is that in some circumstances, especially in Italy, the sense of smell plays a part in the orientation of pigeons, but it cannot by itself explain how pigeons find their homes.

DOES HOMING DEPEND ON MAGNETISM?

In the 1970s and 1980s, the magnetic hypothesis became the most popular among professional researchers (except in Italy, where the smell theory predominated, and still does). The idea was that pigeons could use a magnetic map for homing. It presupposed an exquisitely sensitive magnetic sense in pigeons, by which they could not only detect the compass directions but also sense changes in the Earth's magnetic field from place to place.

In theory, there are two ways in which the earth's magnetic field could give directional information. First, the *strength* of the field varies from the magnetic poles to the equator, being strongest at the poles. Second, the *angle* of the field also varies from poles to equator. Compass needles point downwards at the magnetic poles, and are horizontal at the equator. In between, they dip downwards at angles related to the latitude: more toward the poles, less toward the equator. Hence if the pigeons could detect changes in the strength or the angle of the field, they could sense how far they had traveled towards the magnetic north or south.

On theoretical grounds alone, there are at least three serious problems with this hypothesis. First, the changes in average angle and strength of the field are very small. In the northeastern United States, for example, over a distance of 100 miles in a north–south direction, the average field strength changes by less than 1 percent, and the angle of the field by less than 1°. Second, the earth's magnetic field is far from uniform but varies from place to place depending on the underlying rocks. Some of these "anomalies" are small, a few hundred yards across; others are large, extending over hundreds of miles. In extreme cases, the magnetic field within an anomaly can be up to eight times as strong as the normal magnetic field of the earth. Moreover, the field varies from time to time, both with daily fluctuations and with much larger changes during magnetic storms, due to sun spots. These fluctuations could cause errors from tens to hundreds of miles in reading north–south positions on a magnetic map.[48]

Finally, even if pigeons were sensitive enough to magnetic fields to detect how far northwards or southwards they had been moved, and even if they could also somehow correct for magnetic anomalies and for fluctuations in the field at different times, the earth's magnetic field would give no information on movements in an east–west direction. If a pigeon were taken eastwards or westwards from home, the average angle and strength of the field would be the same as at home, and hence would give no information about the home direction. Yet pigeons can home perfectly well after being taken east or west, or indeed to any point of the compass. So even if pigeons used the strength or angle of the earth's magnetic field to obtain information about north–south movements, there would have to be some other system that gave information about east–west movements. Magnetism could never provide more than a partial explanation of homing, even in principle.

But what if the birds simply have a kind of magnetic compass, rather than a magnetic "map"? As in the case of the "sun-compass," this would not be of much help. A compass cannot by itself give any information about the direction of home.

In spite of these overwhelming theoretical difficulties, the idea that the earth's magnetic field might somehow explain bird navigation was suggested as long ago as 1855, and has cropped up repeatedly ever since.[49] Until the 1970s, this hypothesis met with intense scepticism within the scientific community, if only because many doubted that organisms could detect a magnetic field as weak as the earth's. However, careful experiments in the 1960s in Germany showed convincingly that birds could be affected by magnetic fields. Migratory birds were kept indoors in cages at the time they would normally be migrating. Such birds, not surprisingly, show what researchers call "migratory restlessness," and hop within the cage, trying to move in the general direction in which they would normally migrate. If the magnetic field around such birds was reversed, they would hop in the opposite direction; if it were rotated by 90°, their hopping direction would also change by 90°.[50] By the 1970s, there were several groups of enthusiastic researchers on magnetic orientation. Even the sense of direction of human beings was found to be influenced by weak magnetic fields.[51]

Magnetism, previously dismissed as a wild idea, was now widely welcomed as a scientific explanation for bird navigation, warding off the need for yet wilder ideas. And magnetism is still in general favor, as you can find out for yourself. Raise the subject of bird migration or pigeon homing in general conversation. I predict that many scientifically informed people will say that it has all been explained in terms of magnetism but won't "remember the details."

Here are the details. There are three types of empirical evidence for the influence of magnetism on pigeons' sense of direction, but no evidence that magnetism explains homing. First of all, pigeons are sometimes disoriented when released at places where there are anomalies in the earth's magnetic field; one such place being Iron Mine Hill, Rhode Island.[52] However, they can still home in spite of this initial disorientation. Moreover, only some pigeons are confused by magnetic anomalies. For example, at Iron Mine Hill, birds from a loft in Lincoln, Massachusetts, show a disturbed initial orientation, but birds from a loft in Ithaca, New York, are undisturbed and set off homewards straight away.[53]

Second, pigeons seem to be affected by magnetic storms due to sun spots. Homing speeds in pigeon races tend to be lower in periods of high sun spot activity.[54] Such magnetic storms can affect the direction in which the birds set off, but the average change from the normal homewards direction is small, only a few degrees. And in spite of these initial deflections, pigeons can still find their way home.[55]

Third, pigeons have been deliberately exposed to magnetic fields to see if these confuse them. A wide range of experiments with pigeons and with migrant birds from the 1920s onwards showed no significant effect. Some of the first positive results with pigeons were obtained in 1969 by William Keeton, of Cornell University at Ithaca, New York. He and his colleagues fixed small bar magnets on to the birds' heads or backs. Control pigeons were fitted with small brass bars instead. The magnets had no significant effects on the pigeons' homing on sunny days. But on overcast days, in experiments from 1969–70, the magnets seemed to confuse the birds when they were first released, although they could still home. In subsequent experiments carried out in the early 1970s by other researchers, pigeons had Helmholtz coils fitted to their heads or necks; in the test birds a magnetic field was generated by passing an electric current through the wires of the coil. On sunny days, the magnets had no significant effects. On cloudy days, as Keeton had found, there was some disorientation at first, but the pigeons still homed.[56]

However, these effects of magnets on cloudy days were not

repeatable, even by Keeton.[57] In relation to his early experiments, he himself commented on "the disturbing variability found in the results."[58] From 1971 to 1979, he tried in vain to repeat his initial findings. The negative results of this research were still unpublished when he died in 1980. A posthumous analysis of all his data from thirty-five different experiments on overcast days was published by Bruce Moore in 1988. The effects on initial orientation found in 1969–70 did not show up in the later experiments. Even in the early experiments, the magnets had no significant effect on the ability of the birds to home:

> The birds with the magnets were slightly slower to vanish than those with brass in 1969–70 but were slightly faster in 1971–79. The effects were equal but opposite and neither approached significance. Homing velocities were trivially faster with magnets in both sets of data; but again, neither effect approached significance. Three-fourths of both experimental and control birds reached home on the day of release. . . . Finally, the overall loss rates were identical—26 birds or 9%—in pigeons flown with and without bar magnets.[59]

Pigeons' magnetic sensitivity has also been tested in laboratory experiments. Most of the published results have failed to show any significant effects of magnetic fields, and in addition many other negative studies have remained unpublished.[60] One of the leading investigators in the field, Charles Walcott, has come to the conclusion that: "Given the weight of all this negative evidence, coupled with the circumstantial nature of the positive evidence, it becomes very difficult to believe that the pigeon makes use of magnetic cues for its 'map.' "[61]

The magnetic hypothesis was the last seemingly viable attempt to find a mechanism for homing. Many have clung to it with the tenacity of drowning men clutching straws. Now this hypothesis too has sunk.

Among professional researchers, the currently conventional view is that pigeon homing depends on a complex series of "back-up systems"; or that it is "multifactorial," involving subtle combinations of mechanisms, such as a sun-compass, smell, and magnetism; or that pigeons use a single (unspecified) type of information, but "scan it with several sensory systems."[62] But these scientific-sounding phrases merely disguise a profound ignorance. The orthodox paradigm has broken down.

IS THERE AN UNKNOWN SENSE OF DIRECTION?

The difficulty of explaining bird navigation in conventional scientific terms has been apparent for many years, and today is clearer than ever before. For decades, there has been an undercurrent of speculation about a possible unknown "sense of direction," "faculty of orientation," "sense of location," "sixth sense," and even "extra-sensory perception" or ESP. In the early 1950s, the case for ESP was advocated by several parapsychologists, especially J. B. Rhine[63] and J. G. Pratt,[64] of the Parapsychology Laboratory at Duke University, North Carolina. But the defenders of orthodoxy dismissed such ideas out of hand, confidently asserting that an explanation in terms of normal scientific principles was almost in sight. In the 1950s, the now-discredited sun arc hypothesis seemed the most hopeful. Its leading proponent, G. V. T. Matthews, adopted a magisterially dismissive tone:

> Odd theories calling for "radiations," of an unspecified nature, from the home area frequently crop up in popular literature . . . Rhine (1951) and Pratt (1953, 1956) have suggested that some extrasensory means of orientation is the basis of homing. However . . . no suggestions as to the mode of operation have been forthcoming from the parapsychologists, who really became interested in bird navigation only because the known facts had received no adequate explanation in terms of sensory physiology. This interest has been rejected by Matthews (1956) and there appears to be little activity in this field today. We may also mention here,

and dismiss, vague theories of a special "sense of space" which means nothing and explains less.[65]

Scientific conservatives still cling to the faith that sooner or later an orthodox explanation will be found. But the existence of influences of a kind as yet unknown to science now seems not only possible but probable.

A DIRECT CONNECTION BETWEEN PIGEONS AND THEIR HOMES

I propose that the sense of direction of homing pigeons depends on something rather like an invisible elastic band connecting them to their home, and drawing them back toward it. When they are taken away, this band is stretched. If on their return flight they overshoot their home, as some of the pigeons flying with frosted-glass contact lenses did, this connection serves to pull them back again.

I do not know how this interconnection might work. It might be related to the nonlocal connections implied by modern quantum physics, first pointed out in the Einstein-Podolsky-Rosen paradox. Einstein regarded the nonlocal implications of quantum theory as absurd; he rejected the notion of an instantaneous linkage between two separated quantum systems that had once been together. But, in the form of Bell's Theorem, quantum nonlocality was tested experimentally in 1982 by Alain Aspect, and Einstein turned out to be wrong.

> Assuming one rules out faster-than-light signalling, [this result] implies that once two particles have interacted with one another they remain linked in some way, effectively parts of the same indivisible system. This property of "nonlocality" has sweeping implications. We can think of the Universe as a vast network of interacting particles, and each linkage binds

the participating particles into a single quantum system. . . . Although in practice the complexity of the cosmos is too great for us to notice this subtle connectivity except in special experiments like those devised by Aspect, nevertheless there is a strong holistic flavour to the quantum description of the Universe.[66]

Perhaps the link between the pigeon and its home rests upon such nonlocal quantum phenomena. Perhaps it does not, but rather depends on some other kind of field or interconnection not yet recognized by physics. I simply leave this question open.

Another way of formulating this idea of a connection between a pigeon and its home is through the concepts of modern dynamics. In mathematical models of dynamical systems, systems move within a field-space toward *attractors*.[67] In these terms a homing pigeon could be modeled as a body moving within a vector-field toward the attractor, representing its home or goal.

For the sake of simplicity, I shall take the crudest formulation of this idea, the metaphor of an invisible elastic band between the pigeon and its home. This connection gives pigeons a sense of direction, enabling them to find their way home even when they cannot remember the outward journey, nor see landmarks, nor use a sun-compass, nor smell, nor detect the earth's magnetic field. It enables them to overcome the fiendish confusions imposed upon them by experimenters, including releasing them in heavily overcast weather or at night, shifting their time sense, blocking their nostrils, confusing them with odors, fixing magnets to them, rotating and anaesthetizing them, blinding them with frosted-glass contact lenses, and cutting their nerves.

This band is stretched when the pigeons are taken away from their home. But it should also be stretched under the converse condition: when the home is taken away from the pigeons. This is the basis of the experiment I propose. Instead of the pigeons being taken away from their loft, the loft is taken away from the pigeons. Can the pigeons find their missing home?

The experiment I propose involves a mobile loft. Pigeons are known to be able to home to mobile lofts, and such lofts have been extensively used in the present century for military purpose.

THE MILITARY USE OF MOBILE LOFTS

At the outbreak of the First World War in 1914, the Belgian, French, Italian, and German armies were well equipped with military pigeon services, with many lofts of trained birds, including mobile lofts for use as the army advanced or retreated. The British were entirely unprepared, but after the war broke out, a Carrier Pigeon Service was rapidly built up with the enthusiastic help of amateur pigeon fanciers, organized by Col. A. H. Osman, Officer Commanding Pigeons. Before and after the war, Osman was editor of *The Racing Pigeon* magazine, still the leading British publication in the field. His book *Pigeons in the Great War*[68] gives the definitive account of this remarkable war effort. He tells how the Naval Pigeon Service sent birds out on trawlers engaged in minesweeping work; the pigeons brought back reports to their owners' lofts, who immediately passed them on to the Admiralty. The first reports of an attack on the minesweeping fleet by a zeppelin came by pigeon. Meanwhile, the British Intelligence Corps sent pigeons over German-occupied Belgium, on balloons equipped with clockwork devices for releasing small baskets of birds at intervals. These baskets descended on small parachutes, with requests for Belgians to send information of military relevance. Many actually did so, risking the death penalty imposed by the German authorities. The British Intelligence Service also parachuted spies behind enemy lines, equipped with baskets strapped to their backs containing experienced homing birds wrapped in straw, for sending reports.

Mobile lofts were soon established, and by the end of the war in 1918 the British had more than 150. The American Army Pigeon Service had fifty. Some of these were horse-drawn, others motorized (Figure 3). Pigeons were taken in baskets by motorcycle dispatch riders or on horseback to the troops in the trenches and were used to send messages when radio and other means of communication were impossible. The pigeons even flew to their mobile lofts through heavy artillery fire, and many were commended for their bravery. One British pigeon was awarded the Victoria Cross, and a French one the Légion d'honneur. The American heroine was a blue checkered hen:

> Her last flight was a desperate one on the Argonne, but she bravely got through and delivered her message, although one leg was hanging from the thigh and bleeding profusely. The message was an important one from a platoon in difficulties. Reinforcements saved the situation, and the men of the platoon have cause to bless [her] brave deed.[69]

In the Second World War mobile lofts were used by the British in North Africa, and by the Indian Army Pigeon Service in Burma.[70] The Indian Pigeon Service also developed a "boomerang" flying system whereby the pigeons were trained to find a mobile feeding loft each day, returning to the stationary home loft to perch. The same birds could thus be used to carry messages both ways.[71] A similar system was used successfully by the British Army Pigeon Service in Algeria and Tunisia.[72] And two-way homing systems using mobile lofts are currently being developed in Switzerland, with birds from the Swiss Army Pigeon Service[73] (one of the last military carrier-pigeon services to survive, the other being in China).

Under wartime conditions, the pigeons adapted well to the movements of their lofts. Col. Osman reports that in the First World War "the birds found their homes anywhere." But I have been unable to find out exactly how the mobile lofts were used. Presumably, in most cases, the loft was moved with the birds inside it. Probably they were given an opportunity, if possible, to get used to the new environment before being used for homing. In this case, homing to the mobile loft would not be particularly surprising.

Mobile lofts were also used on ships at sea. In the First World

War, the Italian Navy used them to carry messages from ship to ship, when both the ships were moving. "From distances over 100 km the birds found their own lofts on the boats, which were moving continuously all the time and did not stay at the same place. Even among very similar boats they found their own ship."[74] This is truly amazing, and I only wish more details were available.

FIGURE III

Mobile pigeon lofts used in the First World War. (From Osman and Osman, 1976.)

Motor mobile loft.

German loft, captured and exhibited at London Zoo.

A mobile loft, somewhere in France, camouflaged.

An Experiment
with Mobile Lofts

The experiment I propose involves a mobile pigeon loft, mounted on the back of an old farm trailer. The pigeons in the loft are first trained to home in the normal way, just like ordinary homing birds. Then they are trained to home to the mobile loft. The basic procedure is to take some of the pigeons out of the loft and keep them in a pigeon-carrying basket. The loft is then towed away, still containing some of the pigeons, including the mates and offspring of those removed. Then the birds in the basket are released at the place where the loft used to be. They can see immediately that their home has gone. Can they find it?

If pigeons can find the mobile loft repeatedly, rapidly, over long distances, in arbitrary directions and when the loft is moved downwind (eliminating any possibility of smelling its direction), this would show that there is a direct connection between the pigeons and their home. On the other hand, if the pigeons cannot find the mobile loft, even with the other pigeons inside it, the outcome would unfortunately be inconclusive. It could mean there is no invisible connection between the pigeons and their home. Or it could mean that there is a connection to the home, but moving the loft alone is not enough. More of the home environment might need to be transported, for example by mounting the loft on a ship.

Of relevance in this connection is a report I have received

from a Dutch correspondent, Mr. Egbert Gieskes, of a mobile loft on the river Rhine:

> A Dutch skipper, owner of a riverboat, brought goods from sea vessels in Rotterdam with his boat to Germany or Switzerland. His pigeons were flying every day around his ship during his trip up or down the river Rhine. One day he offered a friend in Rotterdam a basket with three pigeons and said: "Let them free after five days, look what they do and write down the time." Half a day later the pigeons came to their loft in Basel, between a lot of other ships.

This story is not quite as surprising as the use of lofts on ships at sea by the Italian Navy, since the pigeons were familiar with the Rhine and could simply have flown upriver until they found their boat. But it does suggest a potentially simple experiment which could be done with the help of this or any other riverboat skipper who keeps pigeons on the Rhine. Instead of releasing the birds at Rotterdam, at the mouth of the Rhine, where there is only one direction in which they can fly along the river, they could be released roughly half way along the Rhine, say at Koblenz, in Germany. Neither the pigeons nor the person releasing them would know which way the boat was going, to Rotterdam or Basel. If in a series of experiments the pigeons repeatedly went in the right direction and found the loft straightaway, rather than flying up- or downriver at random, this would indicate the existence of an invisible connection between the birds and the loft.

Unless one happens to know a friendly ship's captain, it is simpler to start this line of research with ordinary mobile lofts on land. And the first step is to train the birds to home to the mobile loft over short distances. Pigeons, like people, do not normally expect their home to move. The first time it happens they are very confused, just as most people would be if they went home and found a gap where their house used to be. Even if they could plainly see the house some distance down the road, they would be unlikely to walk straight in as if nothing had happened. But if it kept doing this, they would just get used to it. And so do pigeons.

Training Pigeons to Home to Mobile Lofts

I have trained pigeons to home to mobile lofts in Ireland and in England, and have found they soon get used to their home moving around.

I first had the opportunity to work with a mobile loft in 1973, when the Marquis and Marchioness of Dufferin and Ava kindly offered me the use of their estate at Clandeboye, County Down, in Northern Ireland. I did this research with the help of the agent, Donald Hoy, and the head gamekeeper, Bob Garvin, who looked after the birds from day to day.

We bought a standard two-compartment wooden pigeon loft and mounted it on a farm trailer so that it could be towed around by tractor or by Land Rover. In the summer, twelve mature birds were established in the loft and trained to home in the normal way. Unfortunately, most were lost, shot, or killed by sparrowhawks. We then obtained ten more birds, young ones, and kept them in the other compartment of the loft.

It was not possible to begin experimental research until November, when pigeons are not breeding and are least attached to their homes. By this time only three of the original twelve birds were left, and five of the new ones. It was far from an ideal time for an experiment, but since I was moving to India in the New Year we decided to try training the older birds and see what happened.

The first time we moved the loft, we took it only 150 yards, keeping it within the same meadow. The birds were kept inside the loft, and the three old pigeons were released two days later. For half an hour they flew around the old position of the loft before they began to approach the loft in its new position. It took a further half hour before they landed on the roof, soon taking off again. Finally, an hour and a half after they were released, two of them went in, and were fed. The other one was more timid and spent the night in a nearby tree before entering the loft in the morning.

The next day we moved the loft 50 yards to a new position within the same meadow and released the old birds. They circled over the previous position, but soon landed on the loft in its new position and entered it within 15 minutes, and were fed. The next day the loft was moved to another field, a distance of 300 yards, and the same birds were released. This time they circled the old position only briefly, and entered the loft within 10 minutes. They had clearly got used to the fact that their home could move.

After this brief period of training, we tried the experiment itself. In the morning we put the old birds in a well-ventilated box. We towed the loft, containing the five young pigeons, to a field near Downpatrick, 20 miles south. The test pigeons were then released from their box at the exact place the loft had last been.

I watched them with keen interest. They circled over all four places where the loft had been kept; landed on the ground at those places; perched in nearby trees, and on several occasions disappeared from view for periods of 10 minutes or so, but then returned. After several hours of this fruitless activity, they started following me around, landing at my feet, pecking pathetically at blades of grass. The message was unmistakable: they were hungry. They roosted for the night in a tree, and the next morning they were still in the fields where the loft had been. Again they followed me around. This went on all day, and again they roosted in a tree at night. The next morning I gave in. We brought the loft back, and when we arrived we found the birds were sitting on the ground at the very spot we were planning to put it. They entered the loft within minutes and fed voraciously.

Obviously this preliminary experiment failed to reveal any mysterious navigational powers. But I was not too discouraged. At this season the homing motivation is weak; the training period had been very brief, and the five young pigeons kept in the loft were not related to the test birds, and had been living separately.

I planned to conduct the experiment again during the breeding season when homing motivation is strong. But, alas, this was not to happen. By the time I returned from India on leave some eighteen months later, despite a restocking of the loft, the local sparrowhawks had reduced the population to a mere two birds, and the experiment had to be abandoned.

Another opportunity to set up a mobile loft arose in 1986, thanks to David Hart, at whose estate, Coldham Hall in Suffolk, England, the loft was kept. The birds were looked after by Robbie Robson of Bury St Edmunds, the President of the local Pigeon Racing Association, a fancier with many years of experience. I am very grateful to him for his freely given help.

As at Clandeboye, the mobile loft was made by assembling a standard two-compartment loft from a commercially available kit and mounting it on a farm trailer (Figure 4). We painted broad yellow stripes on the roof, to make it clearly recognizable from the air. The total cost was less than £400. The loft was stocked with young birds kindly donated by local fanciers.

To start with, the loft was kept in a large stable yard behind Coldham Hall. The birds got used to the surrounding area, had experience of homing from up to 50 miles away, and were breeding in the loft. The first time we moved the loft, in July 1987, we took out the eight adult birds, keeping them in a wicker pigeon-carrying basket while we towed the loft across the yard. Six fledglings were left in the loft, together with some chicks. In this and all subsequent tests, the adult birds were then released exactly where the loft had previously been.

As in Ireland, the birds were at first confused when their previously stationary home suddenly moved, even though it was only 100 yards away and plainly visible. They circled repeatedly around the place where the loft had been, from time to time landing on the ground there. But after a quarter of an hour, one of the birds, an unpaired red cock, flew over the loft in its new position. After another quarter of an hour it did so again, and the others followed. All flew over the loft several times in the next half an hour, as if working out a flight path, and the red cock then alighted briefly on the roof. Ten minutes later (80 minutes after release), he went in

FIGURE IV

*The author and Robbie Robson (right) near the mobile loft,
waiting for the pigeons to find it.*

and was fed. After some 10 minutes, he took off and joined the
others, overflying the loft with them, and sometimes landing. But
it took a further 4¹/₂ hours before five others went in, 6 hours after

they were released. The remaining two did not do so until the following day, having spent the night in a nearby oak tree.

The next afternoon we moved the loft another 100 yards, after removing all but one of the adult birds from it. Within 2 minutes of release, the birds started flying over the loft, and all entered it within an hour and a quarter.

We continued training the birds with several more moves in the summer of 1987, and began again in the spring of 1988. We found that when the loft was moved after being left in a particular place for several weeks or months, and especially if the loft was moved to a completely new place, the birds found it quite quickly but were reluctant to land on it or enter it, tending to perch in nearby trees. However, when they were thoroughly used to it moving, their fear diminished. By the summer of 1988, we reached the point where we could take the pigeons out of the loft, tow it a mile or two to a new place, return to the previous position of the loft to release the birds, and then drive back to the loft in its new position, only to find them sitting on the roof waiting to be fed.

All went well until we put the loft near a farm barn about a mile away. The pigeons found the loft but refused to enter it, waiting a week until we moved the loft away from the barn and into a field.

In retrospect, we should have realized that the pigeons were frightened by the barn, or rather by the strange men working there and the movement of farm machinery. After the training program was back on track, we moved the loft another two miles, putting it beside another barn on a neighboring farmer's land. This was a disastrous mistake. The barn was quite heavily used by more strange men with more noisy machines. Although the pigeons soon found the loft, they would not land on it, and started living in the nearby fields, where there was plenty of food, adopting a feral lifestyle.

We moved the loft into one of these fields, but three weeks passed before they could be coaxed back in. This delay, and the need for redomesticating them in the loft prevented any further experiments being done that season. In 1989, we thought, having learned from our mistakes, we would carry out a rapid training program, and then do the big experiment, moving the loft at least 20 miles.

Alas, this was not to happen. During the winter, Robbie Robson fell ill with pigeon lung. Apart from the debilitating symptoms, this meant that he could not work with pigeons any more, since the disease is exacerbated by the dust from feathers. Without Robbie to tend them on a daily basis, the birds reverted to the wild.

HOW TO BEGIN

I have just summarized the state of the art in research with mobile lofts. The field is wide open.

I would strongly advise anyone taking up this experiment to find an experienced pigeon fancier to advise and help them, unless they are already experienced in keeping pigeons. Successful work with pigeons depends on the basic skills of handling, training, and caring for the birds, and on forming a good relationship with them.

In the Practical Details section at the end of the book, I list the addresses of pigeon magazines and organizations where information can be found about local groups of fanciers, loft kits, commercially available pigeon feed, and other practical matters. Young birds can be purchased from local fanciers, or they may well be given. In my experience, most pigeon fanciers are well aware of the unexplained nature of the homing instinct, take a friendly interest in practical research on the subject, and are in any case very helpful to people establishing new lofts.

Once the loft is established, the birds well settled and used to homing in the normal way, they should then be trained to find the mobile loft, starting with short moves. When the pigeons are used to homing to the mobile loft, it can be moved further and further. The further it is moved, the more interesting the results will be.

It is, of course, essential to keep a detailed written record of the setting up of the loft and of the training flights, and in the experiments to take careful notes of the weather conditions, wind direction, exact time of release of the birds, and the time at which they first appear near the mobile loft.

If the pigeons can indeed find their home even after it has been moved far away, say 50 miles, the time they take will be crucial. If they take weeks to find it, this could be the result of random search, and would not therefore provide evidence for a direct connection between the pigeons and their home. But if they reach the loft within an hour or two, they would have had to fly there more or less direct. And if this effect were repeatable at a variety of locations, under conditions when the loft was not upwind, then it would prove the existence of a direct connection between the pigeons and their loft.

Many further questions could be asked. For example: is the connection more to the other pigeons or to the loft itself? To investigate this, the other pigeons could be removed from the loft and kept in one distant place, while the loft itself is taken somewhere else. Do the test pigeons head for the other members of their flock, or for the empty loft? A new field of research would open up.

PETS WHO FIND THEIR OWNERS

If pigeons can indeed find their home and their companions after the loft has been moved far away, then a series of strange but persistent stories about pets will be seen in a new light. There are many accounts of homing pets, as discussed above; but there are also many stories about pets left at home finding owners that have moved. Such stories have been told for centuries. For example, in the sixteenth century, a greyhound named Cesar was said to have followed his master from Switzerland to Paris, setting off three days after his master had left by coach. The dog somehow found his master at the court of King Henri III. In an even more heroic example of canine devotion, we are told that during the First World War a British dog named Prince found his way across the English Channel to his master's side at the battlefront in France.[75]

Most modern cases come to light through reports in local newspapers. For example, when a family was leaving California for a new home in Oklahoma, their Persian cat, Sugar, jumped out of the car, stayed a few days with neighbors, and then disappeared. It turned up a year later at the family's new home in Oklahoma, having traveled well over 1,000 miles through unfamiliar territory.[76] Tony, a mongrel dog belonging to the Doolen family from Aurora, Illinois, was left behind when the family moved more than 200 miles north-east to East Lansing, Michigan, around the southern tip of Lake Michigan:

When the Doolens left Aurora they gave Tony away, but six weeks later he appeared in Lansing, excitedly approached Mr. Doolen on the street, and was recognized. Identity was established by the collar which Mr. Doolen had bought in Aurora and had cut down to Tony's size. A right-angled notch had been cut for an extra hole. Both the Doolen family (of four) and the Aurora family who gave them Tony as a pup recognized the dog, and Tony's behavior confirmed his identity.[77]

There is even a case of a pet pigeon finding its owner, the twelve-year-old son of the county sheriff in Summersville, West Virginia. The racing pigeon, number 167, had stopped in his backyard; the boy fed and cared for it, and it became his pet.

Sometime later the boy was taken to Myers Memorial Hospital at Phillipi, 105 miles away (70 miles by air) for an operation, and the pet pigeon was left behind in Summersville. One dark, snowy night about a week later, the boy heard a fluttering at the window of his hospital room. Calling the nurse, he asked her to raise the window because there was a pigeon outside, and just to humor the lad, she did so. The pigeon came in. The boy recognized his pet bird and asked her to look for the number 167 on its leg, and when she did she found the number as stated.[78]

Such stories naturally arouse much interest, and are widely reported in newspapers and popular magazines. Skeptics inevitably dismiss them as anecdotal, just as they used to dismiss stories about the homing of pets. Experimental research has now confirmed the reality of homing behavior in many animal species, even though it remains unexplained. Likewise, if it can be demonstrated experimentally that pigeons can find homes that have moved, stories about pets finding their owners will have to be taken more seriously.

The biological context for this apparent ability may be the way in which social animals find other members of their group when separated from them. Some observations on wolves by the naturalist William Long seem relevant here:

> In the winter time, when timber wolves commonly run in small packs, a solitary or separated wolf always seems to know where his mates are hunting or idly roving or resting in their day bed. The pack is made up of his family relatives, younger or older, all mothered by the same she-wolf; and by some bond or attraction or silent communication he can go straight to them at any hour of the day or night, though he may not have seen them for a week, and they have wandered over countless miles of wilderness in the interim.[79]

Through long periods of observation and tracking, Long came to the conclusion that this ability could not be explained simply in terms of following habitual paths, or by following scent trails, or by hearing howling or other sounds. For instance, he once found a wounded wolf that had separated from the pack, and lay recovering in a sheltered den for several days while the others ranged widely. Long picked up the trail of the pack while they were hunting, tracked them through the snow, and was close by when they killed a deer.

> They followed, killed and ate in silence, as wolves commonly do, their howling being a thing apart from their hunting. The

wounded wolf was then far away, with miles of densely wooded hills and valleys between him and his pack. . . . When I returned to the deer, to read how the wolves had surprised and killed their game, I noticed the fresh trail of a solitary wolf coming in at right angles to the trail of the hunting pack. It was the limper again . . . I picked up his incoming trail and ran it clear back to the den, from which he had come as straight as if he knew exactly where he was heading. His trail was from eastward; what little air was stirring came from the south; so that it was impossible for his nose to guide him to the meat even if he had been within smelling distance, as he certainly was not. The record in the snow was as plain as any other print, and from it one might reasonably conclude that either the wolves can send forth a silent food call, or else that a solitary wolf may be so in touch with his pack mates that he knows not only where they are, but also, in a general way, what they are doing.[80]

Such connections may be a normal feature of animal societies, even though we have hardly begun to understand how they work. In the following chapter I consider a very different example, termite colonies, in which the individual insects also seem to know where the others are and what they are doing. As in the case of wolves, and of pets that know when their owners are coming home, and of pets that find their owners, and of pigeons that find their lofts, and of homing behavior in general, and of migration, appropriate explanations may lie beyond the current limits of science.

CHAPTER THREE

THE

ORGANIZATION

OF TERMITES

THE TERMITE ORACLE

The social insects—ants, wasps, bees, and termites—have been a source of wonder to people for countless generations. They appear in numerous myths, legends, and fables. In Europe, bees were especially fascinating and were symbolic of death, divination, and regeneration. Some of the oldest goddess images found in Europe are of the queen bee:

> The queen bee, whom all the others serve during their brief lives, was, in the Neolithic, an epiphany of the goddess herself. . . . In Minoan Crete 4,000 years later the goddess and her priestesses, dressed as bees, are shown dancing together on a golden seal found buried with the dead. The hive was her womb—perhaps also an image of the underworld—and later reappears in the beehive tombs of Mycenae. . . . The humming of the bee was actually heard as the "voice" of the goddess, the "sound" of creation. . . . In the Greek Homeric "hymn to Hermes," written down in the eighth cen-

tury BC, the god Apollo speaks of three female seers as three bees or bee-maidens, who, like himself, practiced divination.[1]

In Europe, wasps and hornets were less attractive to the mythic imagination and had a negative image; they were primarily famed for their stings and for their "waspish" character.

Ants, by contrast, attracted a great deal of interest. In Greek mythology they were an attribute of the goddess Demeter. In Celtic lands they were thought of as fairies in their last stage of existence. Ant hills were used for divination and weather forecasting. And in many traditional stories, such as Aesop's fables, ants were noted for their hard work, prudence, orderliness, courtesy, humility, modesty, and uncanny powers of communication.

Most of Europe is too cold for termites and, as the biologist Karl von Frisch has remarked, "the only people that regret that the homes of these interesting creatures are so far away are European biologists."[2] In many tropical regions, they can be extraordinarily destructive, causing houses and other wooden structures suddenly to collapse into piles of dust, having eaten them away from within. But termites are not merely treated as pests: they are held in awe. Among the Dogon of the Sudan, the primal termite mound plays a central role in the story of creation, when the god Amma first makes the body of the earth from a lump of clay:

> This body, lying flat, face upwards, in a line from north to south, is feminine. Its sexual organ is an anthill, and its clitoris a termite hill. Amma, being lonely and desirous of intercourse with this creature, approached it. That was the occasion of the first breach of the order of the universe. . . . At God's approach the termite hill rose up, barring the passage and displaying its masculinity. It was as strong as the organ of the stranger, and intercourse could not take place. But God is all-powerful. He cut down the termite hill and had intercourse with the excised earth. But the original incident was

destined to affect the course of things for ever; from this defective union there was born, instead of the intended twins, a single being, the jackal, symbol of the difficulties of God.[3]

Traditionally in many parts of Africa and Australia, termites are believed to have remarkable powers of communication, and in particular the gift of knowing at a distance. They are used as oracles. For example, among the Azande in West Africa:

The oracle is regarded as very reliable. Azande say that the termites do not listen to all the talk which is going on outside in homesteads and only hear the questions put to them. Of the termites most consulted those called *akedo* and *angba-timongo* are more highly thought of than those called *abio,* which are said to lie often.[4]

The experiment proposed in this chapter also treats termites as an oracle, asking them about themselves. No one knows how their societies are coordinated. Their prodigious organization shows that there must be a sophisticated system of communication within the society. Can this be explained simply in terms of messages passed through smell and other sensory channels, or is their society organized through some kind of field not yet recognized by science?

Before I discuss how this question can be posed experimentally, I need to review the biological background and the various theories proposed to account for the organization of insect societies.

THE BIOLOGICAL BACKGROUND

Termites are often called white ants, but this term is misleading. They are social cockroaches, and probably originated more than 200 million years ago, before the other social insects: the bees, wasps, and ants.[5] Their diet is principally cellulose, which they digest with the help of symbiotic microorganisms and fungi. The more "primitive" species feed directly on the wood in which they nest. The more "advanced" species nest in the soil and forage for dead wood, grass, seeds, and other sources of cellulose. Most species are white and soft-skinned, shunning the light and living in darkness in decaying wood, in nests, and in tunnels. They are blind, except for the winged sexual forms.

Like ants, termite societies have distinct castes, including soldiers specialized for the defense of the colony and versatile workers. But unlike ants, bees, and wasps, where females predominate, termites are partnership societies, containing both male and female workers and soldiers. The queen is accompanied by a king, who may live with her for years at the very heart of the colony.

Once or twice a year, young sexual forms appear and, like winged ants, swarm in enormous numbers. They are a great delicacy for many animals and people. They are generally eaten alive, without their wings, but are said to be particularly delicious roasted.

After the nuptial flight, the survivors shed their wings and form pairs, of which only a few achieve their goal: the construction of a hidden chamber as the nucleus of a new colony. Only then do they achieve sexual maturity and start their life-long matrimony. At first, they tend the brood themselves; later, the brood tends them, and they devote themselves to the task of reproduction.

The larvae of ants, bees, and wasps hatch from their eggs as helpless maggots, and they do not start an active life within the community until they have pupated and metamorphosed. The development of termites is different. Like cockroaches and grasshoppers, they never enter a pupal stage but gradually grow up from moult to moult, increasingly resembling the mature form. Termites work while they are still larvae.

The nests of the more "primitive" species are well hidden, consisting of an apparently irregular system of passages and chambers in wood or soil. The queen may be relatively small and move around. But in more "advanced" species, the nests are much more elaborate, and can be enormous, up to twenty feet high (Figure 5).

The queen is confined to a royal cell, swells up and lays prodigious numbers of eggs; for example in the African species *Macrotermes bellicosus,* the queen may be more than 5 inches long and lay 30,000 eggs a day, living for many years. Colonies may contain several million insects. Some last for centuries, the kings and queens being replaced when they die.[6]

The chambers of termite mounds may extend far into the ground, with networks of subterranean passages and overground tubes leading into the surrounding area where workers collect food. Some desert termites drive boreholes to depths of 100 feet or more to reach water. In many species, the thick, hard outer wall of the mound contains air spaces and ventilation shafts. The nest itself, surrounded by an air space, contains the royal cell, together with many chambers, passages, and the fungus gardens in which the termites cultivate fungi on finely chewed wood.

These structures are built by workers from pellets of soil, first moistened by excrement or saliva and then drying hard. But how do the workers know where to put the pellets?

It is all but impossible to conceive how one colony member can oversee more than a minute fraction of the work or envision in its entirety the plan of such a finished product. Some of the nests require many worker lifetimes to complete, and each new addition must somehow be brought into a proper relationship with the previous parts. The existence of such nests leads inevitably to the conclusion that the workers interact in a very orderly and predictable manner. But how can the workers communicate so effectively over such long periods of time? Also, who has the blueprint of the nest?[7]

Termites raise in an extreme form a problem posed by all animal societies: how are the activities of the individual members coordinated so that the society functions as a whole? The whole seems to be more than the sum of the parts; but in what does this wholeness consist?

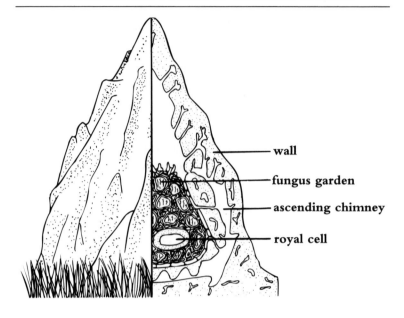

wall

fungus garden

ascending chimney

royal cell

FIGURE V

The nest of the African termite Belliocosotermes natalensis. *The nest is over eight feet high. Around the central area containing the royal cell and the fungus gardens is an elaborate system of air ducts which serve as a ventilation and cooling system. (After Dröscher, 1964, and Noirot, 1970.)*

THE NATURE OF INSECT SOCIETIES: PROGRAMS AND FIELDS

Within biology, insect societies were traditionally thought of organically. The entire society was seen as an organism, or rather a superorganism. Edward O. Wilson, who worked on social insects before he became the leading proponent of sociobiology, described the decline of the superorganism concept as follows:

> During some forty years, from 1911 to about 1950, this concept was a dominant theme in the literature on social insects.

Then, at the seeming peak of its maturity, it faded, and today it is seldom explicitly discussed. Its decline exemplifies the way inspirational, holistic ideas in biology often give rise to experimental, reductionist approaches which supplant them. For the present generation, which is so devoted to the reductionist philosophy, the superorganism concept provided a very appealing mirage. It drew us to a point on the horizon. But, as we worked closer, the mirage dissolved—for the moment at least—leaving us in the midst of unfamiliar terrain, the exploration of which came to demand our undivided attention. . . . There exists among experimentalists a shared faith that characterizes the reductionist spirit in biology that in time all the piecemeal analyses will permit the reconstruction of the whole system in vitro.[8]

However, as Wilson readily admitted: "The total simulation of construction of complex nests from a knowledge of the summed behaviors of the individual insects has not been accomplished and stands as a challenge to both biologists and mathematicians."[9]

The continued failure of the reductionistic approach has led to a recent revival of the superorganism concept.[10] The analysis of the behavior of individual insects is not enough; the holistic properties of the colony have to be admitted. But how should these be investigated?

The most popular approach at present is to try to model these holistic properties on computers, borrowing techniques from those who try to model the activity of the brain. The holistic properties of the colony are then supposed to "emerge" from the interactions of the individual insects, just as the holistic properties of the brain are supposed to emerge from the activity of individual nerve cells.[11] Following this analogy, present-day computer models of insect societies are based on computer models of brains, using "neural networks," "parallel distributed processing," and "cellular automata" techniques.[12] The individual "insects" are programmed with a number of simple responses, and then they are set interacting with their neighbors in accordance with higher-level programs in such a way that social behavior "emerges":

> As in neural systems, behavioral processes will be defined, to some extent, by the kind of connectivity between microscopic parts (ants or neurons). Some kind of collective emergent behavior will be observed as the result of local coupling. . . . In ant societies, such new properties are, for example, nest building, trail formation, or foraging behavior.[13]

This computer modeling is interesting as far as it goes, but it leaves most of the fundamental questions unanswered. What in physical reality corresponds to the overall programs in the computer that coordinate and remember the activity of the individual "insects"? Programs are purposive and mind-like, which is not surprising since they are produced by human minds to fulfill particular purposes. The programs for computer models of insect colonies play the same role as the "soul of the colony" or "group minds" proposed long ago by vitalists, but dismissed by mechanists as "mystical." Computer models do not prove that higher-level mind-like activities "emerge" from mechanistic interactions of nerve cells or insects; they assume this to start with.

Computer models also tell us little about the physical basis of the communication within the colony. Insofar as they assume that interactions among insects depend only on known physical senses such as touch and smell, they may be wrong.

I believe that the most promising approach is to think of the holistic organization of termite colonies in terms of *fields*. The individual insects are coordinated by the social fields, which contain the blueprints for the construction of the colony. Just as the spatial organization of iron filings around a magnet depends on the magnetic field, so may the organization of the termites within the colony depend on a colony field. To make models without taking such fields into account is rather like trying to explain the behavior of iron filings around a magnet ignoring the field, as if the pattern somehow "emerged" from programs within the individual iron particles.

The term "field" was first introduced into science by Michael Faraday in the 1840s, in connection with electricity and magnetism. His key insight was that attention should be focused on the space around a source of energy, rather than on the source of energy itself. In the nineteenth century the field concept was confined to electromagnetism and light. It was extended to gravitation by Einstein in his general theory of relativity in the 1920s. According to Einstein, the entire universe is contained within the universal gravitational field, curved in the vicinity of matter. Moreover, through the development of quantum physics, fields are now thought to underlie all atomic and subatomic structures. Each kind of "particle" is thought of as a quantum of vibratory energy in a field: electrons are vibrations in electron fields, protons are vibrations in proton fields, and so on. Quantum matter fields, electromagnetic fields, and gravitational fields are different in kind, but they all share the common characteristics of fields as regions of influence, with characteristic spatial patterns.

Fields are inherently holistic. They cannot be sliced up into bits, or reduced to some kind of atomistic unit; rather, fundamental particles are now believed to arise from fields. Physics has already been transformed through the extension of field concepts, but this revolution is still in its early stages in biology. A beginning was made in the 1920s, when *morphogenetic fields* were first postulated by several embryologists and developmental biologists to help explain how plants and animals develop. The fields were thought of as invisible blueprints or plans which shaped developing organisms.[14]

The concept of morphogenetic fields is now widely adopted by developmental biologists, and is used to help explain how your arms and legs, for instance, have different shapes in spite of the fact that they contain the same genes and proteins. They differ because your arms developed under the influence of arm morphogenetic fields and your legs under the influence of leg fields. The fields play a formative role in an analogous way to architectural plans. From the same building materials, houses of different shapes can be constructed according to different plans. The plan is not a material constituent of the house; it just shapes the way the materials are put together. Like architectural plans, morphogenetic fields are not reducible to the material components of an organism, nor even to the interactions between these components. The form of the house does not "emerge" from the interactions between its material components; the components interact the way they do because they were put together in accordance with a particular plan, which existed even before the house was built.

The problem is that no one knows what morphogenetic fields are or how they work. Most biologists assume that sooner or later they will be understood in terms of conventional physics and chemistry, but I think they are new kinds of fields, for which I have proposed the term *morphic fields*. In the hypothesis of formative causation, I suggest that the holistic, self-organizing properties of systems at all levels of complexity, from molecules to societies, depend on such fields. Morphic fields are not fixed but evolve. They have a kind of inbuilt memory. This memory depends on the process of morphic resonance, the influence of like upon like through space and time.[15]

However, the purpose of the experiments described below is not to test my own particular version of biological field theory, but rather to test the field approach in general. Are termite societies organized by fields of a kind at present unrecognized by physics? At this stage the question can be left open as to whether such fields are morphic fields, nonlocal quantum fields, or whatever.

THE FIELDS OF TERMITE COLONIES

To suggest that termite colonies are organized by fields is not to deny the importance of normal sensory communication. Like ants, termites are known to communicate in a variety of ways: through sound, through touching each other,[16] through the sharing of food and through smell, using specific chemical signals known as pheromones.[17] In the case of ants, pheromones seem to be the most important means of sensory communication. "In general it appears that the typical ant colony operates with somewhere between 10 and 20 kinds of signals, and most of these are chemical in nature."[18] The most studied of these pheromones are the "alarm" chemicals, which depend on diffusion through the air and typically work over a range of two to three inches,[19] and the pheromones used by the insects to mark their trails, which others then follow.[20]

However, in the making and repairing of nests, the workers do not simply respond to each other but to the physical structures that are already in place. For example, in the building of arches in termite nests, workers first make columns, and then bend them towards each other until the growing ends of two columns meet (Figure 6). How do they do this? The workers on one column cannot see the other; they are blind. There is no evidence that they run back and forth at the base of the columns measuring the distance between them. Moreover, "it is improbable that in the midst of all the confused scampering in the vicinity they can recognize distinct sounds from the column by conduction through the substrate."[21] The sense of smell may play some part, as it does in the communication of ants and other social insects, for example through odor trails, alarm substances, and through the exchange of liquid food. But smell can hardly account for the overall plan of the nest or the relationship of the individual insects to it. They seem to "know" what kind of structure is required; they seem to be responding to a kind of invisible plan. As E. O. Wilson phrased the question, "Who has the blueprint of the nest?" I suggest that this plan is embodied in the organizing field of the colony. This field is not inside the individual insects; rather, they are inside the collective field.

Such a field must cover the entire colony, probably with subfields for particular structures such as tunnels, arches, towers, and fungus gardens. For such fields to play an organizing role, they

FIGURE VI

The construction of an arch by workers of the termite species Macrotermes natalensis. *The columns are built up of pellets of mud and excrement, carried by the insects in their mouths. (After von Frisch, 1975.)*

must be capable of permeating the material structures of the colony, passing through the walls and chambers. Just as a magnetic field can pass through material structures, so can the colony field. This ability to pass through material barriers would enable the field to organize separated groups of termites even in the absence of normal sensory communication between them.

So the question is: are the nest-building activities of termites still harmoniously coordinated even when sensory communication is blocked by means of a barrier? The analogy with the magnetic field is again helpful: if the arrangement of particles of iron in lines of force depended only on the particles directly contacting their neighbors, then the magnetic field pattern would be disrupted by a mechanical barrier, for example by a sheet of paper. But in fact the line patterns cross the barrier because they depend on a field to which this barrier is permeable.

Termites are known to be sensitive to magnetic fields, the most spectacular illustration being the compass termites of Australia, which orient their nests with their narrow sides pointing north and south, minimizing heating by the sun in the middle of the day. Termites have also been shown in the laboratory to respond to very weak alternating electrical and magnetic fields.[22]

Moreover, Günther Becker in Berlin has demonstrated in laboratory experiments that termites can influence each other by what he calls a "biofield," which could be electrical in nature. From a captive colony of the species *Heterotermes indicola* he took groups of about 500 workers and soldiers and put each group in a rectangular polystyrene container with wood and damp vermiculite. He then placed eight of these containers in adjacent rows, four in each row, with gaps of 1 cm between containers. After several days the termites began to make galleries up the corners of the containers. But they did not do so equally at all corners but only on outside corners; almost no galleries were made on the sides of the container adjacent to other containers. This pattern resembles that found in actual termite nests, where the galleries are not built in the center but on the periphery, extending outwards toward potential food and water supplies. In a typical experiment, the total length of galleries on the outward-facing sides of the containers was 1,899 cm, compared with only 80 cm on the inward-facing sides. In other experiments, Becker found that when isolated containers were kept more than 10 cm away from any other container, there was more gallery building activity than in a similar container kept close to others; when other groups were close by, gallery building was suppressed. Somehow groups of termites influenced each other in a way that fell off with distance.

In another experiment, Becker arranged 16 containers in a 4 × 4 pattern, so that there were four containers on each side of the array, and four central containers. Again the great majority of gallery building took place on the outward-facing sides (Figure 7), while on the inward-facing sides and in the four central containers there was very little gallery building (a total of only 43 cm per day compared with 590 cm per day on the outward-facing sides). Becker interpreted these results in terms of the "biofield" inhibiting gallery building in the central part of the field.

Termites still inhibited gallery building among their neighbors when additional barriers were placed between the containers, such as polystyrene foam and tall plates of glass. Becker thought that these additional separating layers prevented the conduction of temperature and vibration, and also excluded any possible chemical influences, but the biofield could pass through polystyrene and glass. However, when thin aluminum plates or fibreboard covered with a silver-containing paint were slotted between the containers, the biofield effect was abolished; termites built galleries on inner-facing walls as well as outer-facing ones, and also in the inner four containers of the 4×4 array. The aluminium barriers and silver paint screened out electrical fields, and Becker suggested that the "biofield" was probably an alternating low-energy electric field produced by the termites themselves.

However, given that electrical and magnetic fields can influence the building activities of termites, such fields are unlikely to be able

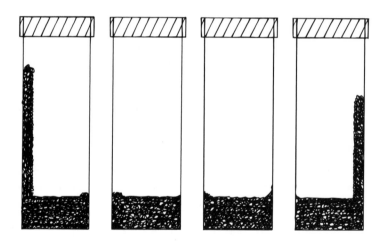

FIGURE VII

The construction of vertical galleries by termites of the species Heterotermes indicola *kept in captivity in plastic containers containing an inert building material, vermiculite. All containers contained equal numbers of termites. Gallery building was suppressed on the walls adjacent to other containers. This influence passing from container to container was mediated by fields. (After Becker, 1977.)*

to provide the blueprint for the termite nest. For how could a specific pattern be established in the electromagnetic field to begin with? Another, more mysterious, kind of field seems likely to be involved as well.

Experiments carried out by the South African naturalist Eugène Marais suggest that such a field exists. In the 1920s, Marais made a fascinating series of observations of the way workers of a *Eutermes* species repaired large breaches he made in their mounds. The workers started repairing the breach from every side, each carrying a grain of earth which it coated with its sticky saliva and glued into place. The workers on different sides of the breach did not come into contact with each other, and could not see each other, being blind. Nevertheless, the structures built out from the different sides joined together correctly. The repair activity seemed to be coordinated by some overall organizing structure, which Marais attributed to the group soul, and I prefer to think of as a morphic field.

> Take a steel plate a few feet wider and higher than the termitary. Drive it right through the centre of the breach you have made, in such a way that you divide the wound and the termitary into two separate parts. One section of the community can never be in touch with the other, and one of the sections will be separated from the queen's cell. The builders on one side of the breach know nothing of those on the other side. In spite of this the termites build a similar arch or tower on each side of the plate. When eventually you withdraw the plate, the two halves match perfectly after the dividing cut has been repaired. We cannot escape the ultimate conclusion that somewhere there exists a preconceived plan which the termites merely execute. Where is the soul, the psyche, in which this preconception exists? . . . Where does each worker obtain his part of the general design? We can drive in the steel plate and then make a breach on either side and still the termites build identical structures on each side.[23]

Marais's experiments imply the existence of an organizing field which, unlike the gallery-inhibiting field investigated by Becker, was not blocked by a metal plate, and was therefore unlikely to be electrical in nature.

Marais took this research further, with results that imply that the organizing field is intimately linked to the queen, and that the death of the queen causes an immediate disruption of the entire field:

> While the termites are carrying on their work of restoration on either side of the steel plate, dig a furrow enabling you to reach the queen's cell, disturbing the nest as little as possible. Expose the queen and destroy her. Immediately the whole community ceases work on either side of the plate. We can separate the termites from the queen for months by means of this plate, yet in spite of this they continue working systematically while she is alive in her cell; destroy or remove her, however, and their activity is at an end.[24]

As far as I know, no one has ever tried to repeat Marais's experiments. The reductionist climate of modern biology is inhospitable to Marais's approach, and his work has been ignored by professional researchers. But in my opinion his findings provide the most promising starting point for a new wave of research on the organization of insect societies.

Proposed
Experiments

1. First of all, it seems important to repeat Marais's experiment with the steel plate. Are the repair activities on both sides of the plate as well coordinated as Marais claimed?

This experiment will not be feasible for those who live in the colder parts of the world, unless they are prepared to establish termite colonies indoors. But in tropical countries where termites are common, to repeat Marais's work may be relatively easy. The termite mounds come free; the only expense is the steel sheet. However, I imagine that driving a large steel plate into a termitary may be difficult. And it may well be difficult to withdraw it again without major disruption after the termites have healed the breach. Marais gives no details, so the only way to find out is to try for one's self.

If the repair activities of the termites on the two sides of the barrier are as well coordinated as Marais claimed, then many further experiments become possible. Do other kinds of barriers give similar results to steel? Can termites pass sound signals across these barriers? What happens to the pattern of activity of termites on one side of the barrier if the repair activities on the other side are prevented or disturbed? And so on.

2. Do disturbances to the queen affect the entire colony very rapidly, as Marais claimed? In the passage quoted above, he called this effect "immediate." In another instance he describes how he was observing the queen of a very large colony, having opened up the royal cell, when a piece of hard

clay fell on to the queen, dealing her a hard blow. The workers within the royal cell immediately ceased work and wandered round in aimless groups. He then visited far outlying parts of the nest many yards away:

> Even in the farthest parts all work had ceased. The large soldiers and workers gathered in different parts of the nest. There appeared to be a tendency to collect in groups. There was not the least doubt that the shock to the queen was felt in the outermost parts of the termitary within a few minutes.[25]

Possibly these dramatic effects spread through the colony as a result of sound signals, by insects releasing alarm pheromones in relays or by some other conventional means. But they could also have been transmitted very rapidly through the organizing field, if such a field exists. In the latter case, the transmission might well still take place through barriers blocking the passage of sounds or smells.

Rather than killing the queen, or dropping hard objects on her, this experiment could be done simply by removing the queen, or by anaesthetizing her and the termites all around her. The activities of termites in distant parts of the colony would be closely observed while this was happening. The speed at which the disturbance spread could then be worked out. If almost immediate, then alarm pheromones could be ruled out, but sound would still be a possibility. It would be hard to rule out sound by means of barriers, because it would be difficult to prove that no sounds could have gone through or around them, unless sensitive microphones had been installed at various places within the colony to monitor the sound signals.

A better way of investigating the possible transmission of influences through fields would be to have a part of the colony within a portable structure that can be taken away from the main part of the colony. This could be, for example, a metal box previously placed near the nest in which the termites had constructed nest structures, or a metal box containing food to which the workers habitually went to feed. If such a box were removed, the termites within it would still be part of the colony but now deprived of all normal physical connections with the queen and other colony members. No doubt the mere fact of removing the box would disturb the insects within it, but if they were kept under continuous observation, it might be possible to observe dramatic changes in their behavior when the queen was disturbed or anaesthetized in the main part of the nest.

3. Similar experiments should also be possible with ants, which are relatively easy to keep in captivity. Working with ants is feasible almost anywhere in the world. Multichambered containers for ant colonies are commercially available for less than £25, and there are well-established methods for making one's own out of very inexpensive materials such as plaster of Paris, plastic tubs, and plain glass. Details are given in the Practical Details section at the end of this book.

The simplest design consists of a two-chambered colony in which the two parts are connected by a plastic tube. The two parts can be separated simply by disconnecting the tube and plugging the holes. Then one part of the colony could be taken away to another room while the other part, containing the queen, is left where it is. The part containing the queen is then disturbed, for example, by vibration, or smoke, or by an anaesthetic such as ether, and the separated part is closely observed to see if any changes occur that could indicate an "action at a distance."

In all such experiments it is important to work "blind" as much as possible. For example, the person observing the chamber that has been moved should not know exactly when the queen-containing chamber is going to be disturbed. If dramatic changes in the ants' behavior are seen, and the time at which they happen is later found to correspond exactly to the time of disturbance, this would provide good evidence for a transmission of influence. Further experiments could then be carried out taking the separated chamber further and further away to see how far this influence can spread. Tests could then be made to see if the influence could be blocked by metal or other kinds of barriers, and so on. As soon as any repeatable effect is found, the nature of the organizing field can be investigated progressively.

CONCLUSIONS

TO PART ONE

All the experiments proposed in the previous chapters concern possible connections of a kind at present unknown to science— connections between pets and their owners, between pigeons and their homes, and between the insects within a termite colony. All of them have enormous implications. If pets have unseen connections with people, what about connections between people and wild animals, taken for granted in shamanic traditions for millennia? And if this kind of interspecies communication occurs, then what about unknown kinds of connections *within* species?

If pigeons depend on a hitherto unrecognized connection for their homing, then so may many other species with homing abilities. Similar powers may play an important part in the migrations of birds, fish, mammals, insects, and other animals. Even the human sense of direction, so well developed among hunter–gatherers and nomadic peoples, may have a similar component.

If termites are coordinated through a field that links the individual members of the colony, could similar systems of interconnection be at work in other social animals, including schools of fish and flocks of birds? Could this help to explain how such groups can turn as one, without the individuals bumping into each other? How might such unknown fields of communication be related to the "group minds" of herds of animals, and of human groups? Could they also be related to the bonds between pets and their owners?

Possibly the experiments will give no evidence for the existence of such connections. Then the sceptical opinions of scientific conservatives will be reinforced. The failure of attempts to find new kinds of connections will lend support to the conventional faith that all possible kinds of interconnection between organisms are already known, and that they are in principle, if not in practice, fully explicable in terms of the established laws of physics and chemistry.

But it is also possible that in some or all cases the existence of new kinds of connection will be established by experiment. What would these results suggest?

First and foremost, it is obvious that the success of any or all of these experiments would lead to a reinterpretation of homing, migration, the sense of space, bonding, social organization, and communication in general. Biology would be revolutionized. What about physics? If the results of experiments in biology lead to the need to postulate new kinds of fields or connections, how might these be related to the known principles of physics?

One possibility is that there are many different kinds of undiscovered fields. The connections between pets and their owners, between pigeons and their homes, and between the members of termite colonies may be entirely different phenomena and have nothing in common. They may each depend on a new kind of field or physical connection capable of acting at a distance, but apart from this very general resemblance, the fields or connections involved may be completely different.

I prefer the more economical hypothesis that these phenomena may all be related. They may all be manifestations of a new kind of field which embraces the separated parts of an organic system and connects them together (Figure 8). I would think of such fields as morphic fields. Others may prefer to propose different names for

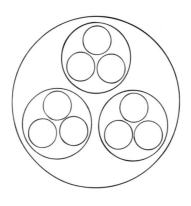

FIGURE VIII

TOP: *Successive levels of organization in self-organizing systems. Each level of organization depends on a characteristic morphic field. In the chemical realm, for example, the outer circle represents the morphic field of a crystal, the circles within it the fields of molecules, and the circles within those the fields of atoms, which in turn contain the fields of subatomic particles. In the case of social animals, the outer circle represent the morphic field of the social group, the inner circles the individual animals within it, and the circles within those the animals' organs.*

BOTTOM: *A representation of the way the morphic field of a social group is stretched when one or more members of the group are separated from the others. This field acts as an unseen bond or connection between the separated members of the group. These general principles would apply to the connections between a pet and its absent owner; between a pigeon and its flock-mates in the loft; and between separated members of a termite colony.*

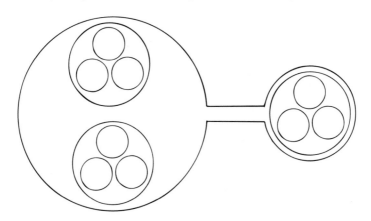

these fields, which could in general be called biological fields or life fields.

Sooner or later, any new kind of field must in some way be related to the known fields of physics, even if this relationship becomes clear only in the light of a unified field theory of the future. Such a unified theory would need to be far larger in scope than any attempted to date, since institutional physics has so far ignored the possibility of any fundamentally new field phenomena in the realm of life.

But while so little is actually understood, and while the outcome of these experiments remains to be seen, these questions are entirely open.

PART TWO

THE

EXTENDED

MIND

CONTRACTED

AND EXTENDED

MINDS

We know very little about the nature of our minds. They are the basis of all our experience, all our mental and social life, but we do not know what they are. Nor do we know their extent. The traditional view, found all over the world, is that conscious human life is part of a far larger animate reality. The soul is not confined to the head but extends throughout and around the body. It is linked to the ancestors; connected with the life of animals, plants, the earth, and the heavens; it can travel out of the body in dreams, in trance, and at death; and it can communicate with a vast realm of spirits—of ancestors, animals, nature spirits, beings such as elves and fairies, elementals, demons, gods and goddesses, angels and saints. Christian versions of this traditional understanding were prevalent all over Europe throughout the Middle Ages, and still survive in rural societies, for example in Ireland.

By contrast, for more than three hundred years the dominant theory in the West has been that minds are located inside heads. This theory was first propounded by Descartes in the seventeenth century. Descartes denied the old belief that the rational mind was part of a larger soul, mainly unconscious, pervading and animating the entire body. Instead, he supposed that the body was an inani-mate machine. Animals and plants were machines too, and so was the entire universe. In his theory, the realm of soul shrank from nature into man alone, and then in the human body contracted yet further into a small region of the brain, which Descartes identified as the pineal gland. The conventional modern theory is essentially the same, except for the fact that the supposed seat of the soul has moved a couple of inches, into the cerebral cortex.

This model of the contracted mind, confining the soul to the brain, is shared by both sides in the familiar, long-standing debate between dualists and materialists. Descartes himself, the prototypic Cartesian dualist, regarded the mind and brain as fundamentally different in nature, yet interacting within the brain in an unknown manner. By contrast, materialists reject this dualist conception of a "ghost in the machine," and believe that the mind is nothing but an aspect of the mechanistic functioning of the brain, or else that it is an inexplicable "epiphenomenon," rather like a shadow, of the brain's physical activity. But although these rigorous materialist views are espoused by some philosophers and ideologists, dualism is far more prevalent in our culture, and is usually regarded as common sense.

In the older imagery of popular science, the machinery was controlled by little men inside the brain (Figure 9). In more up-to-date images, the machinery has been modernized, but the homun-culi are still there, if only implicitly. For example, in a current exhibit at the Natural History Museum in London entitled "Con-trolling Your Actions," you can find out how you work by looking through a Perspex window in the forehead of a model man. Inside is the cockpit of a modern jet plane, with banks of dials and com-puterized flight controls. There are two empty seats, presumably for you, the ghostly pilot, and your co-pilot in the other hemi-sphere. The currently fashionable computer metaphor for the brain is no different: if the brain is the hardware, and habits and skills the software, then you are the phantom programmer.

How many people really think of themselves as machines? Even ardent materialist philosophers and mechanistic scientists do not

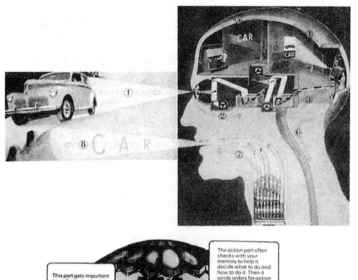

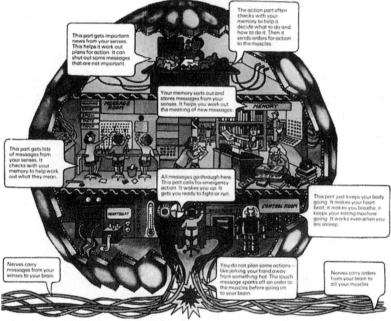

FIGURE IX

Little people inside the brain.

(A) The image in a book of popular science entitled The
Secret of Life: The Human Machine and How It Works
*(Kahn, 1949). The caption reads: ''This is what takes place
in the eye, brain and larynx when we see a motor car,
recognize it as such, and pronounce the word 'car.' ''*

*(B) The image from a contemporary children's book, widely
used in British schools, called* How Your Body Works
(Hindley and Rawson, 1988).

seem to take this belief very seriously, at least in relation to themselves and their loved ones. In personal as opposed to official life, most people still retain to varying degrees the older and broader perspective of their ancestors. First, the soul is often thought to pervade more of the body than the brain. And second, the soul is widely believed to participate in extended psychic and spiritual realms, stretching far beyond the confines of the body.

In Hindu, Buddhist, and other traditional psychologies, there are several animating centers within the body, the *chakras,* each with their own characteristic properties. In the West too, various psychic centers have traditionally been recognized, in addition to the head. For example, people often talk of "gut feelings." And although from a mechanistic point of view the heart is just a pump, expressions such as "heartfelt thanks," "heartless behavior," and "warm-heartedness" obviously refer to more than a blood-pumping mechanism. So does the heart as a symbol of love. Indeed, our ancestors believed that the focus of psychic life was in the heart, not in the brain. The heart was more than a center of emotion, love, and compassion: it was a center of thought and imagination, just as it still is for many traditional peoples today, including the Tibetans. Think, for example, of the phrases still used in Christian liturgy: in the Magnificat: "He hath scattered the proud in the imagination of their hearts"; and in the Collect for Purity in the Book of Common Prayer: "Almighty God, unto whom all hearts be open, all desires known and from whom no secrets are hid, cleanse the thoughts of our hearts by the inspiration of thy Holy Spirit."

The old sense of the psyche as extended *beyond* the limits of the body is also widespread in our culture. This is implied in common turns of phrase such as "your ears must have been burning yesterday, because we were talking about you." It is also implied by telepathy and other psychic phenomena. In Britain, the U.S.A., and other Western countries, surveys have repeatedly shown that a substantial majority of the population believes in their occurrence, and more than 50 percent claim to have personal experience of such phenomena.[1]

Such experiences and beliefs do not make sense if the mind is confined to the brain, nor if all communication depends on the known principles of physics. For these reasons, defenders of mechanistic orthodoxy often assert that since "paranormal" phenomena cannot be scientifically explained, they cannot exist. Belief in them is regarded as a superstition, to be eradicated through scientific education.

What started as a radical philosophy has now become the orthodox doctrine of our culture, picked up in childhood and thereafter taken for granted. According to the classic studies of Jean Piaget on the mental development of European children, by the age of about ten or eleven most children have learned what he calls the "correct" view, namely that thoughts are located inside the head.[2] By contrast, younger children believe that they travel outside their bodies when they dream; that they are not separate from the living world around them, but participate in it; that thoughts are in the mouth, breath, and air; and that words and thoughts can have magical effects at a distance. In short, young European children show the animistic attitudes that are found in traditional cultures all over the world, and which were prevalent in our own culture until the mechanistic revolution.

However, the Cartesian theory of an immaterial mind within a machine-like brain ran into serious problems from the outset. By equating the soul with the rational mind, Descartes denied the bodily and unconscious aspects of the psyche, previously taken for granted. Ever since Descartes, the unconscious psyche has had to be reinvented.[3] For example, in 1851, the German physician C. G. Carus wrote a treatise on the unconscious which began as follows:

> The key to an understanding of the nature of the conscious life of the soul lies in the sphere of the unconscious. . . . The life of the psyche may be compared to a great, continuously circling river which is illuminated in only one small area by the sun.[4]

Through Sigmund Freud's work, the recognition of the unconscious became widespread among psychotherapists; and in Carl Jung's notion of the collective unconscious, the psyche is no longer confined to individual minds, but shared by everyone. It includes a kind of collective memory in which individuals participate unconsciously.

There has also been a growing awareness in the West of Indian, Buddhist, and Chinese traditions, all of which offer a richer understanding of the relation of the psyche to the body than the mechanistic theory. And through explorations of the effects of psychedelic drugs, the visionary practices of shamans, and oriental techniques of meditation, the existence of other dimensions of consciousness has become a matter of personal experience for many Westerners.

Thus, although the confining of the mind to the head of a machine-like body is still orthodox in mechanistic science and medicine, it coexists with survivals of an earlier and broader understanding of the psyche. It is also subject to the articulate and sophisticated challenges posed by Jungian and transpersonal psychology, psychical research and parapsychology, mystical and visionary traditions, and holistic forms of medicine and healing.

The experiments proposed in this part of this book explore the possibility that the mind is indeed extended beyond the brain, as most people throughout most of human history have believed. Although the theory of the contracted mind is a central feature of the mechanistic paradigm, it is not an unchallengeable dogma to which science is forever committed. It should be regarded as a scientifically testable, and potentially refutable, hypothesis. The following experiments are designed to put it to the test.

CHAPTER FOUR

THE SENSE OF

OF BEING STARED AT

Do Minds Reach Out From Brains?

When we see things, where are they? Are their images inside our brains? Or are they outside us, just where they seem to be? The conventional scientific assumption is that they are inside the brain. But this theory may be radically wrong. Our images may be outside us. Vision may involve a two-way process, an inward movement of light and an outward projection of mental images.

For example, as you read this page, light rays pass from the page to your eyes, forming an inverted image on the retina. This image is detected by light-sensitive cells, causing nerve impulses to pass up the optic nerves, leading to complex electrochemical patterns of activity in the brain. All this has been investigated in detail by the techniques of neurophysiology. But now comes the mystery. You somehow become aware of the image of the page. You experience it outside you, in front of your face. But from a conventional scientific point of view, this experience is illusory. In reality, the image is supposed to be inside you, together with the rest of your mental activity.

Traditional peoples all over the world take a different view. They trust their own experience. Vision reaches out from the body. As well as light coming into the eyes, seeing goes out through the eyes. Young children in our own culture think the same.[1] But by the age of around eleven, they learn that thoughts and perceptions are not outside them, but inside their heads.[2] Thus theory triumphs over experience, and a metaphysical doctrine is accepted as an objective fact. From the "educated" point of view, young children, like primitive and uneducated people, are confused. They fail to distinguish between internal and external, subject and object, which should be sharply separated.

Consider for a moment the possibility that young children and traditional peoples may not be as confused as we usually suppose. Try a simple thought experiment. Allow yourself to trust rather than distrust your own direct experience. Permit yourself to think that your perceptions of all the things you see around you are indeed *around* you. Your image of this page, for example, is just where it seems to be, in front of you.

This idea is so staggeringly simple that it is hard to grasp. Although in perfect accord with immediate experience, it undermines everything we have been brought up to believe about the nature of the mind, the interiority of subjective experience, and the separation of subject and object. Instead of the usual assumption that vision involves a one-way process, it implies a two-way process. As well as light coming into the eyes, images and perceptions are projected outwards through our eyes into the world around us.

Our perceptions are mental constructions, involving the interpretive activity of our minds. But while they are images in our minds, at the same time they are outside our bodies. If they are both within the mind and outside the body, then the mind must extend beyond the body. Our minds reach out to touch everything we see. If we look at distant stars, then our minds stretch out over astronomical distances to touch these heavenly bodies. Subject and object are indeed confused. Through our perceptions, the environment is brought within us, but we also extend outwards into the environment.

In normal perception, that which we perceive—for example, this printed page—and our perceptual image of it coincide; they are in the same place. In illusions and hallucinations, the images do not coincide with things outside us, but may nevertheless involve a similar process of projection, an outward movement of images. (I return to a discussion of this question in chapter 5 in connection with phantom limbs.)

This idea of the extended mind may sound like playing with words, a mere intellectual exercise. Or it may sound like an illegitimate confusion of philosophical categories that ought to be kept separate: the physical realm on the one hand, and the phenomenological or subjective realm on the other. But it is not just a matter of words or of philosophy. The extended mind may have measurable effects. If our minds reach out and "touch" what we are looking at, then we may affect what we look at, just by looking at it. If we look at another person, for example, we may *affect* him or her by doing so.

Is there any evidence that people can tell when they are being looked at by someone, even when they cannot see the person looking at them? For example, can people tell when they are being stared at from behind? As soon as we ask this question, we realize that there is a great deal of anecdotal evidence suggesting that this is the case. Many people have had the experience of feeling that they are being looked at, and on turning around find that they really are. Conversely, many people have stared at other people's backs, for example in a lecture theater, and watched them become restless and then turn round.

THE POWER OF LOOKS

The sense of being stared at is very well known. In informal surveys in Europe and America, I have found that about 80 percent of the people I asked claimed to have experienced it themselves. It is also taken for granted in countless works of fiction, as in the phrase "she felt his eyes boring into the back of her neck." It is explicitly described by novelists such as Tolstoy, Dostoevsky, Anatole France, Victor Hugo, Aldous Huxley, D. H. Lawrence, J. Cowper Powys, Thomas Mann, and J. B. Priestley.[3] Here is an example from a short story by Arthur Conan Doyle, the creator of Sherlock Holmes:

> The man interests me as a psychological study. At breakfast this morning I suddenly had that vague feeling of uneasiness which comes over some people when closely stared at, and, quickly looking up, I met his eyes bent upon me with an intensity which amounted to ferocity, though their expression instantly softened as he made some conventional remark upon the weather. Curiously enough, Harton says that he had a very similar experience yesterday upon deck.[4]

The veteran British psychical researcher Renée Haynes has described some of her own informal observations on the subject as follows:

> The impulse to turn is not equally strong in everyone, and there will be instances—as with waiters—when it is probably atrophied, ignored, or directly resisted. A little mild experimentation though, say at a boring lecture or in a crowded canteen, will show that in the majority of cases to gaze intently at the back of someone's head will produce a fidget and an uneasy turn and glance. This can also be done with sleeping cats and dogs—not to mention children who may be roused thus more humanely than with a cold sponge—and with birds in the garden.[5]

Probably the effects of looks play an important part in the relationship of people with their pets, and not only may the animals respond to people in this way but people may also respond to animals. In *The Call of the Wild,* Jack London, a keen literary observer of canine behavior, described a particularly intimate situation involving the dog Buck:

> He would lie by the hour, eager, alert, at Thornton's feet.
> . . . Or, as chance might have it, he would lie farther away,
> to the side or rear, watching the outlines of the man and the
> occasional movements of his body. And often, such was the
> communion in which they lived, the strength of Buck's gaze
> would draw John Thornton's head around, and he would
> return the gaze, without speech, his heart shining out of his
> eyes as Buck's heart shone out.[6]

There is also much anecdotal evidence for the influence of looks within other species, in the wild. Here, for example, is a naturalist's account of the power of looks in foxes:

> I have spent hours by different dens, and have repeatedly
> witnessed what seemed to be excellent discipline; but I have
> never heard a vixen utter a growl or cry of warning of any
> kind. For hours at a stretch the cubs romp lustily in the
> afternoon sunshine, some stalking imaginary mice and grass-
> hoppers, others challenging their mates to mock fights or
> mock hunting; and the most striking feature of the exercise,
> after you have become familiar with the fascinating little
> creatures, is that the old vixen, who lies apart where she can
> overlook the play and the neighborhood, seems to have the
> family under control at every instant, though never a word is
> uttered. Now and then when a cub's capers lead him too far
> from the den, the vixen lifts up her head to look at him
> intently; and somehow that look has the same effect as the
> she-wolf's silent call; it stops the cub as if she had sent a cry or
> a messenger after him. If that happened once you might over-
> look it as a matter of mere chance; but it happens again and
> again, and always in the same challenging way. The eager cub
> suddenly checks himself, turns as if he had heard a command,

catches the vixen's look, and back he comes like a trained dog to the whistle.[7]

In the 1980s, when I realized the enormous theoretical implications of this phenomenon, I tried to find out what empirical investigations had been carried out. I was amazed to discover that there had been very few. I gave a lecture on this subject to the British Society for Psychical Research in London, hoping to find that some of the members might have some special knowledge of experimental research on the effects of staring, but once again drew a blank, although the redoubtable Renée Haynes had, as usual, a rich stock of anecdotes on the subject. I also discussed the matter with various parapsychologists in the United States, and found that none had yet worked on the subject, or paid it any attention.[8] Searching back in the scientific archives, I have been able to find only six papers on the subject in the last hundred years, and two of these are unpublished. Orthodox psychologists have ignored the phenomenon, which is not surprising, given its "paranormal" quality. What is more surprising is that parapsychologists have ignored it too. Most books on parapsychology do not even mention it. The fact that even parapsychologists have neglected this phenomenon is in itself very interesting, and suggests that there has been an extraordinary "blind spot," almost amounting to an unconscious taboo. Why should this be so? Perhaps because the sense of being stared at is so closely related to beliefs that modern people would like to dismiss as superstitious, in particular the "Evil Eye."

THE EVIL EYE

A belief in influences transmitted through the eyes is found in practically all traditional societies.[9] In its negative form, it is the Evil Eye, the eye of envy that blights what it looks upon. "He that hasteth to be rich hath an evil eye," in the words of the Book of Proverbs.[10] Young children, cattle, crops, houses, cars, and indeed anything capable of being envied are supposed to be affected by it. It causes ill-health and misfortune. This is why so many precautions are taken against it, for example in the form of amulets. In modern Greece, these usually take the form of a single blue eye, the eye of Horus, a direct descendant of one of the magical talismans of ancient Egypt.[11] This eye is a major feature of the Great Seal of the United States, and can be seen on any dollar bill (Figure 10).

The original meaning of the word "fascination" referred to this power of casting a spell through looking, a usage which survives in relation to snakes immobilizing their prey through their gaze. In the mythology of ancient Greece, the glare of the snake-haired Medusa turned men to stone; the mask of the Medusa, in the form of the Gorgon's Head, represented the terrifying power of the goddess Athena.[12]

Here are Sir Francis Bacon's reflections on the subject of fascination in his essay "On Envy," published in 1625:

> There be none of the affections which have been noted to fascinate or to bewitch, but love and envy; they both have vehement wishes, they frame themselves readily into imaginations and suggestions, and they come easily into the eye, especially upon the presence of the objects which are the points that conduce to fascination, if any such there be. We

FIGURE X

The radiant Eye of Horus on the Great Seal of the United States, as shown on every dollar bill.

see likewise the Scripture calleth envy an evil eye. . . . So that still, there seemeth to be acknowledged, in the act of envy, an ejaculation, or irradiation of the eye. Nay some have been so curious, as to note, that the times when the stroke, or percussion, of an envious eye doth most hurt, are, when the party envied is beheld in glory; for that sets an edge upon envy.[13]

The very word "envy" is derived from the Latin *invidia*, from the verb *invidere,* to see intensively. But although envy is the most frequent emotion associated with the Evil Eye, other negative emotions such as anger are also believed to affect people through the eyes, as in the familiar phrase "she looked daggers at him." In our own society, staring at people is generally considered to be rude, and tends to cause discomfort or provoke aggressive responses.

Some people's looks are supposed to be more harmful than others, and those with "the Eye" are feared as bringers of misfortune. Such beliefs were also widespread in medieval England, and witches were often accused of "overlooking" children or domestic animals that fell ill for no apparent reason. As the Egyptologist Sir Wallis Budge put it:

Various conclusions have been arrived at by those who have studied the why and wherefore of the Evil Eye, but in no part of the world is it doubted that its influence exists and the belief in it is beyond all doubt primeval and universal. Moreover, every language, both ancient and modern, contains a word or expression which is the equivalent of "Evil Eye."[14]

The positive effects of looks, especially loving looks, are also widely acknowledged. In India, for instance, many people visit holy men and women for their *darshan,* literally their look, which is believed to confer great blessings. Perhaps an unconscious survival of the same kind of belief is involved in the popular desire to see in person the Queen of England, the President of the United

States, the Pope, pop stars, or other luminaries. Although they can be watched more conveniently on television, there is something about their real presence which has an enormous appeal, motivating people to spend hours waiting in crowds to catch a glimpse of them—and, better still, to be seen by them. ("The Queen waved at me!") Such people are, as we say, "in the public eye."

In summary, the idea that influences can pass out of the eye is practically universal. This implies an implicit belief in the extension of the mind, capable of influencing what is seen. The ignoring or denying of this possibility by conventional science is not based on a careful consideration of the evidence; the subject is rarely even discussed. Rather, it is based on the conventional assumption that the mind is inside the brain: the theory of the contracted mind. The possibility that there might really be some mysterious effect of looks is simply out of the question; it is dismissed as a matter of principle.

Obviously the question cannot be settled on the basis of scientists' prejudices, nor by popular belief, nor by the piling up of anecdotal evidence, nor by theoretical disputes about the nature of the mind. The only way forward is by means of appropriate experiments.

THE SCIENTIFIC BACKGROUND

The first discussion of the feeling of being stared at in the scientific literature appeared in *Science* in 1898, in an article by E. B. Titchener, a pioneering scientific psychologist at Cornell University, in New York State:

> Every year I find a certain proportion of students, in my junior classes, who are firmly persuaded that they can "feel" that they are being stared at from behind, and a smaller proportion who believe that, by persistent gazing at the back of the neck, they have the power of making a person seated in front of them turn round and look them in the face.[15]

Titchener was confident that there must be a rational explanation, admitting of no mysterious influences. It is worth reading his account in detail, since exactly the same kind of explanation is still given by present-day sceptics:

> The psychology of the matter is as follows:
> 1. We are all of us more or less nervous about our backs. If you observe a seated audience before it has become absorbed in the music or lecture for which it came together, you will notice that a great many women are continually placing their hands to their heads, smoothing and patting their hair, and every now and then glancing at their shoulders or over their shoulders to their backs; while any of the men will frequently glance at or over their shoulders, and make patting and brushing movements with their hand upon lapel and coat-collar. . . .
> 2. Since it is the presence of the audience, of people seated behind one, that touches off the movements described above, it is natural that these movements should in many cases be so extended so as to involve an actual turning of the head and sweeping of the eyes over the back of the room or hall. . . . Observe that all this is entirely independent of any gaze or stare coming from behind.
> 3. Now, movement in an unmoved field—whether the field be that of sight, or hearing, or touch, or any other—is one of the strongest known stimuli to the passive attention. . . . Hence if I, A, am seated in the back part of the room, and B moves head or hand within my field of regard, my eyes are fatally and irresistably drawn to B. Let B continue the movement by looking round, and, of course, I am staring at him. There are, in all probability, several people staring at him, in the same way and for the same reason, at various parts of the room; and it is an accident whether he catch my eye or another's. Someone's eye he almost certainly will catch. These accidents evidently play into the hands of a theory of personal attraction and telepathic influence.

THE SENSE OF BEING STARED AT

4. Everything is now explained, except the feeling that B experiences at the back of the neck. This feeling is made up, upon its sensation side, simply of strain and pressure sensations which, in part, are normally present in the region (sensations from skin, muscle, tendon and joint), but are now brought into unusual prominence by the direction of attention upon them, and, in part, are aroused by the attitude of attention itself. . . . The "feeling of Must" in the present case is no more mysterious than the "feeling of Must" that prompts us to shift our position in a chair, when the distribution of pressures has become uncomfortable, or to turn our better ear to the sound that we wish particularly to observe. 5. In conclusion, I may state that I have tested this interpretation of the "feeling of being stared at," at various times, in a series of laboratory experiments conducted with persons who declared themselves either particularly susceptible to the stare or peculiarly capable of "making people turn round." As regards such capacity and susceptibility the experiments have invariably given a negative result; in other words the interpretation offered has been confirmed. If the scientific reader objects that this result might have been foreseen, and that the experiments were, therefore, a waste of time, I can only reply that they seem to me to have their justification in the breaking-down of a superstition, which has deep and widespread roots in the popular consciousness. No scientifically-minded psychologist believes in telepathy. At the same time, the disproof of it in a given case may start a student upon the straight scientific path, and the time spent may thus be repaid to science a hundredfold.[16]

Whereas to some on the "straight scientific path" this may still seem convincing, others will notice that Titchener assumes what he sets out to prove. The scenario he describes could well have included a mysterious influence of staring. And his experimental disproof of the phenomenon, of which he gives no details, might have other explanations. For example, his subjects might have been put off by his sceptical attitude, or too self-conscious to perform well when tested by him under artificial conditions in a laboratory.

Herein lies the greatest problem for investigating this phenomenon by experiment. The "feeling of being stared at" may work under natural conditions in an unconscious manner. Under artificial conditions, in experimental trials, trying to decide consciously whether one is being stared at may be difficult without practice. Moreover, in real life there are a variety of feelings associated with the act of staring, such as anger, envy, or sexual attraction. If in experimental tests all motivation is removed, save for scientific curiosity, then the effects may be much weakened.

The second investigation of this phenomenon was published in 1913 by J. E. Coover. Following up Titchener's investigations, he found that 75 percent of the students in his junior classes at Stanford believed in the reality of the feeling of being stared at. He then carried out experimental tests with ten subjects. Each was looked at from behind by the experimenter in a series of 100 trials. The experimenter (Coover himself or an assistant) either looked at the subject or looked away, in a random sequence, indicating when the trial began by a tapping noise. The subject then said whether or not she was being looked at, and then said how certain she felt in her guess. The overall result showed that subjects were right only 50.2 percent of the time, not significantly better than the chance level of 50 percent. Nevertheless, when the subjects said they were very certain of their guess, they were correct 67 percent of the time; when they were less certain the results were around or slightly below chance levels. Coover dismissed this aspect of his own findings. He concluded that although the belief in the feeling of being stared at is common, "experiment shows it to be groundless."[17]

That was more or less the end of the matter for nearly half a century, until the subject was raised again in 1959 by J. J. Poortman in the *Journal of the Society for Psychical Research.*[18] He described tests he carried out in Holland, with a woman friend looking at him—a member of the City Council of The Hague who had told him that "she resorted to staring at a person whom she saw at a gathering and to whom she wished to speak." He followed the same method as Coover; in a sequence of 89 trials, carried out on several different days, the lady councillor looked or did not look in a random sequence, and recorded whether Poortman said yes or no. He was right 59.6 percent of the time, as against the chance expectation of 50 percent. This result was statistically significant.[19]

Nearly twenty more years elapsed before the next investigation, in 1978, by a graduate student at Edinburgh University, Donald Peterson. In a series of experiments with eighteen different subjects, he found that they could tell when they were being looked at significantly more often than by chance.[20]

In 1983, an Australian undergraduate, Linda Williams at the University of Adelaide, did a project in which the starer and subject were in different rooms, 60 feet apart. The subject was viewed by the starer by means of closed-circuit television. In a sequence of trials, each 12 seconds long, the starer either saw the subject on the screen, or the screen went blank. (The TV screen was programmed to switch on and off in a random sequence, but the video camera was on all the time. The subject was informed when each new 12-second trial began by an electronic bleep.) The overall results from twenty-eight subjects showed a small but statistically significant positive effect; they did better than chance at knowing when they were being looked at on television.[21]

The most technically sophisticated tests of this ability were carried out in the late 1980s at the Mind Science Foundation, in San Antonio, Texas, by William Braud, Sperry Andrews, and colleagues. They too used closed-circuit television. The subjects were asked to sit calmly in a room for 20 minutes, with the video camera running continuously, thinking of whatever they liked. The starers watched them on a television screen in the viewing room, in a different block of the laboratory building. In contrast to all the earlier experiments, the subjects were not asked to make any guesses about when they were being looked at. Rather, their unconscious bodily responses were monitored, through measuring their basal skin resistance by means of electrodes on their left hand.

Changes in this resistance, as in lie-detector tests, give a sensitive measurement of the unconscious activity of the sympathetic nervous system. In a series of 30-second trials, with rest periods in between, the subject was either being looked at or not, in a random sequence. The results showed significant differences in the skin resistance when the subjects were being looked at, even though they were not conscious of it.[22]

In summary, although there has been remarkably little research on this subject, the available evidence suggests that there is indeed a sense of being stared at, although it does not show up very impressively under artificial conditions.

MY OWN INVESTIGATIONS

I have carried out two kinds of experiments. In the first kind, conducted with several groups in Europe and America, four people volunteered to act as subjects, and sat at one end of the room with their backs toward the rest of the group, who were seated at the other end of the room. In each trial, one of the four was looked at by the rest of the group; the other three were not. At the beginning of each trial I held up a card with the name of the person to be looked at, determined according to a random sequence. At the end of each 20-second trial, all four subjects wrote down whether they thought they were being looked at or not. The results showed that most people, under these conditions, performed little or no better than chance. But in the course of these experiments, I encountered two people who were nearly always right, scoring way above chance levels.

As it happened, both these people were very confident in their ability. The first, a young woman in Amsterdam, said that she had practiced this ability as a child with her brothers and sisters, as a game, and felt sure she would still be able to do it. The second, a young man in California, told me after the experiment that he was under the influence of MDMA, a psychoactive drug commonly known as Ecstasy, and felt a heightened sensitivity as a result.

My second experimental procedure involved immediate feedback: the receiver was told after each guess whether it was right or wrong. Otherwise the procedure was similar to that of most previous researchers: lookers and subjects worked in pairs, with a random sequence of trials. The details are given in the following section.

In these experiments a few people did remarkably well: they were right most of the time. Two of those who did best in my tests were from Eastern Europe; perhaps years living under repressive communist regimes had given them a strong motivation to sense when they were being watched. Most people's results were close to chance levels, but there was nevertheless a significant statistical tendency for people to do better than chance. The overall results from ten different experiments (involving more than 120 subjects) were 1,858 correct guesses as against 1,638 incorrect guesses; in other words 53.1 percent of the guesses were correct, 3.1 percent above the chance level of 50 percent. This result is highly significant statistically.[23]

These results confirm the positive findings of other researchers, summarized above. But they also confirm that most people do not perform very impressively under artificial conditions. The overall results are better than chance, but not much better. The challenge is to find people able to do well under these artificial conditions; my preliminary results show that this may well be possible. Some kinds of people may be especially sensitive anyway. Paranoids, for example, might be exceptionally talented in this respect, but they would probably be paranoid about the experiment itself. People who practice subtle all-round awareness through martial arts such as aikido might be particularly good subjects.

Possible

Experiments

I begin with the outline of the simple experimental procedure that I have tried out extensively. This was designed with a three-fold purpose. First, it is kept as simple as possible, so that it is easy to do. It can be done with groups of people splitting up into pairs; for example, in workshops, classes, or seminars. It can also be done by pairs of people at home or anywhere else; it needs no laboratory, nor any apparatus other than a pencil, a sheet of paper, and a coin—and the coin can be recycled indefinitely. In fact the experiment is free.

Second, it enables people who are unusually talented to be identified, and thus opens the way to more detailed experiments.

Third, it enables people who do not do particularly well to practice and find out if they improve with experience. It may be possible to train oneself to perform well under these artificial conditions. And this too would open the way for further research.

In these experiments people work in pairs, one sitting with his or her back to the other. In a series of trials, in a random sequence, the looker either looks at the back of the subject for 20 seconds, or looks away and thinks of something else for 20 seconds. The random sequence is determined by tossing a coin before each trial: heads means look; tails means don't look. The looker indicates when a trial is beginning by a tap, click, or bleep, and the subject then guesses whether he or she is being looked at or not. Uniform mechanical clicks or electronic bleeps are better than taps because they rule out the possibility of subtle cues being transmitted through the strengths of the taps. The looker records the result, and then tells the subject whether the answer was correct or not. The looker then tosses a coin to determine what to do in the next trial. And so on. The procedure is quite fast, and an average speed of two trials per minute is easy to achieve. The results are recorded on a simple score sheet, as shown in the Practical Details section at the end of this book.

I have found it best to keep test periods fairly short, up to about 20 minutes, during which time forty or more trials can be done. For statistical analysis, at least ten separate test periods are desirable, either with the same pair of people or with different pairs of people.[24]

The procedure described above has already been successfully tried out as a school science project by a 13-year-old in California. Michael Mastrandrea, an eighth-grade student, carried out 480 tests with twenty-four different people. In each case he was the looker. He used an electronic bleeper to indicate when each trial began. The overall results showed that the subjects were correct 55.2 percent of the time, and this positive result was statistically significant.[25]

For those who do not perform particularly well in initial tests, it is good to practice, doing 15- to 20-minute test sessions whenever convenient. This makes it possible for a learning process akin to biofeedback to occur, whereby various subtle sensations or methods of visualization are tried out in the attempt to find an effective way of telling when one is being looked at. If there is a tendency to improve with experience, it should be revealed by a rising proportion of correct guesses in successive sessions.

If and when sensitive subjects have been identified, many further questions can then be asked. Here are some straightforward examples:

1. How much difference does the looker make? Are some people much more effective as lookers than others?

2. Does the sense of being stared at still happen when the person is looked at through a window? Does it still show up when looked at from a distance, for example through binoculars? By experiments of this type, it should be possible to rule out the possibility that in tests in the same room the subjects are being influenced by subtle cues, such as the sound of the looker moving his or her head. If the effect still shows up at a distance, or through soundproof windows, then this would greatly strengthen the evidence for a direct influence of looks.

3. Does the ability show up when the subject's reflection is looked at in a mirror?

4. Does this ability show up when the subject is looked at using closed-circuit TV, with the looker and the subject in separate rooms, or even separate buildings? The results from Adelaide and San Antonio, summarized above, suggest that it does.

5. If it works on closed-circuit TV, then what about actual transmissions? In this case, the effect of distance can be tested over hundreds, or even thousands, of miles, using satellite links. If preliminary experiments show it works on TV, then live experiments could be done involving millions of viewers. Here is one possible design for a TV show. Four sensitive subjects are kept in separate rooms in front of TV cameras that are running continuously. Then, in a series of trials, viewers see one subject at a time in a randomized sequence. At the end of each trial, all four subjects press a button indicating yes or no. Viewers see a scoreboard on which the number of right and wrong scores for each subject is registered. The sequence of trials need take no more than about 10 minutes. A computerized statistical analysis could be available almost immediately, and the rest of the program could consist of a discussion of the results and their implications.

If sensitive subjects are available, there would probably be no problem getting this kind of experiment performed, as I have found by talking to TV producers in Europe and America. Such experiments would make good television and arouse much popular interest.

6. How closely related is the sense of being stared at to telepathy? Does looking at someone have a greater effect than just thinking about them without looking? The way to find out is by experiment. For example, the experiment can be modified to include a third condition, in which the lookers *think* of the receivers but do not look at them. In other words, there would be three kinds of trial in random sequence: looking; not looking and not thinking; not looking but thinking. My own guess is that the effects of looking will be greater than just thinking.

These are only a few of the experiments possible with sensitive subjects, but these examples suffice to show that this could rapidly grow into a fertile field of research. The field is wide open, and the implications are mind-boggling.[26]

CHAPTER FIVE

THE REALITY

OF PHANTOM

LIMBS

THE EXPERIENCE OF PHANTOM LIMBS

When people lose a flesh-and-blood limb, they do not usually lose the sense of its presence. It feels as if it is really there, even though it is no longer materially real. What kind of reality does the phantom have?

In the United States alone, there are more than 300,000 people who have had arms or legs amputated, including some 26,000 war veterans.[1] Nearly all of them have phantom limbs, and although some of the phantoms tend to shrink with time, they rarely disappear altogether. Indeed, in many cases they remain all too vivid an experience: the seat of various pains. Phantom pains really hurt.

Just after amputation, the phantom can feel so real that people who have had their leg removed can easily forget it isn't there. Some even have falls when they try to stand up and walk off. Others "involuntarily put down their hand to scratch their departed foot."[2] People who have recently had an arm amputated often try to reach out and pick up the telephone or other objects.

In addition to a sense of its shape, position, and movement, amputees generally experience various feelings within the missing limb, such as itching, warmth, and twisting. Phantom limbs can generally be moved at will, and they also move in coordination with the rest of the body. Indeed, they are felt to be part of the body. Even when a phantom foot seems to be dangling in the air several inches below the stump, it is still experienced as belonging to the rest of the body, and moves appropriately with the other limbs and torso.[3] One of the curious features of phantom limbs, consistent with their ghostly nature, is that they can be pushed through solid objects such as beds and tables.

I have received dozens of vivid and fascinating accounts of the experience of phantoms from amputees. Some came in response to an article I wrote in 1991 in the *Bulletin of the Institute of Noetic Sciences;* others were from readers of *Veterans of Foreign Wars* magazine, following a notice in the April 1993 issue, kindly placed there on my behalf by Dr. Dixie McReynolds. The following is from Mr. Herman Berg, a veteran who had a leg amputated in 1970:

You get used to the various feelings, itchings and outright pangs of "pain" as time goes by although occasional cursing occurs. Amputation also turns you into a real reliable weather forecaster. You always know there will be a change of one kind or another.

I can always feel the missing leg as being there. At first it seemed to hang through the bed or stick straight up. That's stopped, but it's always there. Then again for days you won't notice the feelings. I can will my mind and have it bend my toes, knees or whatever. I can feel them move through the cut nerves, but there is a jangled or short-circuit feeling while I'm doing this, weird!

Right now as I scribble this I'm sitting at my desk in my shorts and the missing limb is where it should be over the chair in position, and the toes have some feeling too.

Many amputees are troubled by pain from time to time, but unfortunately there is usually not a lot that doctors can do, unless the pain is in the stump rather than in the phantom limb itself. Partially effective methods include meditation and biofeedback practices.[4] Some amputees seek consolation in drink or drugs. But many learn to live with the problem with great courage and cheerfulness. Mr. Leo Unger, for example, had both feet badly damaged by a landmine when fighting in Europe in November 1944. Both his legs were amputated below the knees.

From the very first day I have always had the feeling that my legs and feet were still in place. Early on I had severe phantom pains that felt like balls of fire going down my limbs and off my toes. After 20 years I seldom got that feeling, but I do often feel the bones in my feet were just broken, just as they were when I was wounded. I learned to elevate my legs and the feeling ceases.

For a number of years I was claimsman for Country Mutual Insurance Company, the Illinois Farm Bureau Company, and when some farm hand lost a leg in a corn picker, as many did with the old type pickers, I'd go visit them quite soon after their accident. The first thing I'd tell them was, "Thank God you are an amputee, not a cripple." Then I'd take off my leg that matched his loss, show him what a properly prepared stump should look like, and give him a chance to see how nicely the limb shop can fix one up. Not infrequently they would write to my company to say my home demonstration helped them more than the claim settlement money.

I can't run but I farmed, put in milking parlors, sold insurance, worked as a claims adjuster, and have done nearly everything I wanted for nearly 50 years.

OTHER KINDS OF PHANTOM

Other parts of the body can also give rise to phantoms after their loss, including the nose, testes, tongue, breasts, penis, bladder, and rectum.[5] Sometimes this phantom is agreeable, as it is for some women after the amputation of a breast:

> The painless phantom breast after a mastectomy, in which the nipple is the most vivid part, is usually a pleasant experience because the phantom breast seems to fill out the padded brassiere and feels extremely real. However, pain in the phantom breast becomes distressing.[6]

Likewise, phantom penises can either be pleasant or unpleasant. Some men have painless phantom erections. Some have phantom orgasms. But some suffer pain. One man with severe pain in his phantom penis "was constantly aware of his pain and had often to check a pressing desire to reach out into interpersonal space and squeeze the apparition's tip for relief."[7]

Other kinds of phantom can feel just as real. Some people with phantom bladders keep complaining of a full bladder and even feel that they are urinating. Some with phantom rectums "may actually feel that they are passing gas or feces."[8]

One of the commonest kinds of loss, and hence of phantom, is of fingers and toes. For example, in a case reported in the *British Medical Journal,* a sailor who accidentally cut off his right index finger was plagued by it for decades because the phantom finger seemed to be rigidly extended, just as it was when it was cut off. Whenever he moved his hand to his face, for example to scratch his nose or to eat, he was afraid the phantom finger would poke his eye out. Although he knew this was impossible, the feeling was irresistible.[9]

EXCEPTIONS

Although the loss of body parts usually results in phantoms, there are a few exceptions. Some people who lost parts of their body when they were babies or toddlers have no phantoms. Nor do sufferers from leprosy who lose fingers and toes as the disease progresses. Unlike the loss of fingers and toes through accidents or amputations, the loss through leprosy is gradual, taking up to ten years or even more. The loss of the digit is preceded by the gradual degeneration of the nerves and absence of all sensation in the affected parts. Leprosy is not painful, and degenerating parts of the body can be quite seriously injured and infected without the sufferer paying much attention. As a result, these structures sometimes need to be amputated. But immediately after surgery on the stumps, or after the amputation of the hand or foot, something surprising happens. Even if the leprous digits were lost twenty or thirty years earlier with no phantoms, vivid phantoms of the lost fingers or toes suddenly appear![10]

One of the early theories of phantoms was that they were a kind of memory. It was therefore assumed that they would be absent from people born without a limb (aplasia), for example as a result of their mother having taken thalidomide, a now-banned tranquillizer, during pregnancy. But although most people born without a limb do not seem to have phantoms, between 10 and 20 percent do.[11] Some born without hands experience the presence of fingers, which can be bent. Others born with shortened arms feel their arms to be longer than they really are. For example, a man whose right forearm was almost completely missing, with his hand attached to his elbow, subjectively felt the defective arm to be as long as his normal one.[12] Unlike most phantoms following amputation, the phantoms of congenitally absent limbs are hardly ever painful.[13]

PHANTOMS OF INTACT ORGANS

Phantoms can arise when *sensation* in the limb is lost rather than the limb itself. In some motorcycle accidents, for example, the rider is thrown on to the road in such a way that the shoulder is wrenched forwards, ripping the arm nerves from the spinal cord (a condition known as brachial plexus avulsion). A phantom arm appears. This usually occupies the now-useless true arm and is coordinated with it. But when the victim's eyes are closed, the phantom can separate from the flesh-and-blood arm and take on an independent existence. And although the material arm is no longer capable of responding to stimulation, the phantom arm is often extremely painful. Sometimes the true arm is amputated in an attempt to relieve the pain. Unfortunately for the sufferer, the phantom arm and pain usually remain.[14]

Phantoms are also experienced by paraplegics, who have a broken spinal cord. Sufferers are partially paralysed, with no feeling or control of the body below the break. Nevertheless, they often experience phantom legs and other organs, including phantom genitals.

Usually the phantoms of paraplegics move in coordination with their bodies, especially when their eyes are open, but some paraplegics complain that they cannot keep their phantoms still. For example, their phantom legs may keep on making continuous cycling movements, even when they are lying immobile on a bed.[15]

Just as phantoms can appear when nerves are severed, so they can when nerves are anaesthetized. This phenomenon often occurs in the context of orthopedic surgery. Many patients given a local anaesthetic in the spinal cord experience phantom legs, the proportion depending on the location of the anaesthetic. In one study, 10 percent of those with epidural anaesthesia had phantoms, while 55 percent had them with subarachnoid anaesthesia.[16] The phantom legs are usually partly flexed, and thus when patients are lying on their backs they rise up in the air above the actual legs.

Likewise, after anaesthesia of the nerves running to the arms in the brachial plexus, phantom arms appear. They do so even more frequently than phantom legs, with around 90 percent of patients experiencing them.[17] In one experimental study, patients about to have surgery on their arm or hand were asked to provide a running commentary about their anaesthetized arm, and also to track its position with the other arm. Within 20–40 minutes of the injection, the phantom had appeared:

With the eyes closed, the subject reported that the anesthetized arm felt normal in terms of its position in space; using his tracking arm, he generally showed it to be at the side of the body and bent at the elbow, or above the abdomen or lower chest. The real arm at this time lay flat beside the body. Sometimes the experimenter moved the anesthetized arm slowly until the lower arm and hand were beside the head. When the subject opened his eyes, he was astonished to find the discrepancy between the real anesthetized arm and the perceived arm. The reality of the phantom arm to the subjects was unequivocal. . . . Some of them failed to recognize their real arm when it was raised above their head, and stared in disbelief at it and the place where they perceived it to be.[18]

When the subjects looked at the anaesthetized limb and realized the discrepancy, in most cases the phantom rapidly moved into the real limb, fusing with it. But the phantom limb soon moved up into its previous position when they closed their eyes again. However, in a few heavily anaesthetized subjects fusion did not occur even when their eyes were open: "the phantom retained its ghostly station in spite of repeated instructions to look at the real arm and concentrate on it."[19]

Most anaesthetized patients with phantom arms found that they could move them voluntarily, particularly flexing and extending the hands and moving the phantom fingers. As the anaesthetic wore off, the phantoms disappeared as sensation and active movement returned to the limb.[20]

Phantom arms can also be produced experimentally by putting a pressure-cuff, of the type used by doctors in measuring blood pressure, around the upper arm. If the inflated pressure-cuff is left on long enough, the arm becomes insensate. And if subjects cannot see their arm, within 30–40 minutes most experience it in a different position from the real arm. The phantom disappears when the pressure-cuff is removed and sensation returns.[21]

THE ANIMATION OF ARTIFICIAL LIMBS

Just as phantoms that arise when nerves are severed or anaesthetized can separate from a flesh-and-blood limb and then fuse with it again, so phantoms can fuse with artificial limbs. In fact they play a very important part in people's adaptation to mechanical devices replacing lost parts (called "prostheses" in medical terminology). As one researcher expressed it: "The phantom functions in the control and appreciation of the movements of the artificial limb. At first unrelated, the two come together, arrive at a spatial coincidence, and the lifeless appendage is animated by the living phantom."[22] As another put it tersely: "the phantom usually fits the prosthesis as a hand fits a glove."[23]

In amputees who do not wear artificial limbs, there is a tendency for the phantom to shorten. But the use of prostheses counteracts this shortening, and can even lead a shrunken phantom to grow again. The following example is from Weir Mitchell, the American Civil War surgeon who first introduced the term "phantom" into the medical literature:

In about one-third of the leg cases, and in one-half of the arm amputations, the patient asserts that the foot or hand, as the case may be, is felt to be nearer to the trunk than is the extremity of the other limb. . . . Sometimes it continues to approach the trunk until it touches the stump, or lies seemingly in its interior—the shadow within the substance. . . . Now, we may conceive that if, for motor purposes, we substitute for the lost limb an artificial member which does not possess feeling, the sense of sight will soon refer, in our consciousness, the hand or foot to its old position. Exactly this is described as occurring by two observant and acute persons who have lost legs. One of them, who sees in his business capacity hundreds of the amputated every year, assures me that his experience is a common one. He lost his leg at the age of eleven, and remembers that the foot by degrees approached, and at last reached, the knee. When he began to wear an artificial leg it resumed, in time, its old position, and he is never, at present, aware of the leg as shortened, unless for some reason he talks and thinks of the stump and of the missing leg.[24]

People who wear artificial limbs usually take them off to go to bed, and then the phantom may become very painful. William Warner, an American veteran who lost his right leg above the knee in Italy in 1944, described it as follows:

I get it so bad at times I am unable to sleep. I have talked to a few doctors, but there isn't much they can do. Sometimes at night I get up and put my limb on and walk around, it helps some. As soon as I take it off, it starts in again.

Oliver Sacks has described a similar case where the amputee explicitly thought of the phantom in two different ways: the good phantom that animated his prosthesis and allowed him to walk, and the bad phantom that hurt when the prosthesis was off at night. Sacks comments: "With this patient, with all patients, is not *use* all-important, in dispelling a 'bad' (or passive, or pathological) phantom, if it exists; and in keeping the 'good' phantom . . . alive, active, and well?"[25]

THE FOLKLORE OF PHANTOMS

Amputations have been happening for thousands of years. Hand prints with finger amputations have been found in caves in France and Spain from as long as 36,000 years ago, and artificial arms have been found buried with mummies in Egypt.[26] Organs have no doubt been lost as a result of accidents and fights from time immemorial. Amputations have also been carried out as a punishment, as in the ancient Hebrew code of retaliation, "eye for eye, tooth for tooth, hand for hand, foot for foot,"[27] and in the traditional Islamic penalty for stealing, namely chopping off the right arm of the thief. Thus we can be fairly certain that phantoms and phantom pain are by no means new phenomena, and that they have been known and discussed for millennia. We might also expect that folklore about phantoms has been passed on from generation to generation.

The sensitivity of amputees to the weather is legendary, and here the folklore is continually reinforced by experience. "Involuntary movements of the absent toes or fingers are frequent, and in very many persons are unfailing precursors of an east wind."[28] It would be relatively easy to investigate empirically the accuracy of such weather forecasts, and also to find out whether they could be fully explained in terms of temperature, humidity, barometric pressure, and other straightforward physical factors.

Other aspects of the folklore are harder to check, but no less fascinating. One which crops up recurrently recalls the traditional magical belief that a separated part of the body still retains a connection with it by a kind of action at a distance, or nonlocal connection. I first encountered this way of thinking when I was living in Malaysia. One day when I was staying in a Malay village, a *kampong,* I was cutting my fingernails, throwing the parings into a nearby bush. When my hosts saw this, they were horrified. They explained that an enemy might pick them up and use them to harm me by witchcraft. They were amazed that I did not know that bad things done to my nail clippings could cause bad things to happen to me.

I subsequently discovered that such beliefs are very widespread, and are one of the fundamental principles of sympathetic magic, expressed concisely by the anthropologist James Frazer as follows: "Things which have once been in contact with each other continue to act on each other at a distance after the physical contact has been severed."[29] One of the most intriguing features of quantum theory is that the principle of nonlocality—as expressed in the Einstein—Podolsky—Rosen paradox and Bell's Theorem—makes much the same point about physical processes in the subatomic realm.

In relation to phantom limbs, the belief is that the fate of the severed limb continues to affect the person of whom it was once part. Stories I have received from readers of *Veterans of Foreign Wars* magazine show that this tradition is still alive and well. One man, William Craddock, described how he originally came to know of this belief from his father, who worked as a boiler keeper and maintenance man at a hospital in Jacksonville, Illinois:

> In the 1940s I used to stop at the boiler house on my way home from school. One day my father had something wrapped in cloth on the workbench and he tried to hide it as I came in. I could see the cloth had blood on it, and when I asked my father what it was he said never mind. He later told me it was an amputated limb and that he had just wrapped it to make sure nothing was bent in an unnatural way. He told me he knew a man who was suffering great pain with an amputated arm and they finally dug it up and straightened his fingers. His pain left.

And here is another story, of a man who had a finger amputated and preserved in a jar:

> The man was OK for several years. Then he went back to his doctor, who had amputated his finger, complaining of a feeling of extreme cold in the missing finger. The doctor wanted to know where the jar with the missing finger in it was. The man told him it was in his mother's heated basement where it had always been. The doctor told the man to call his mother and check the jar. The mother didn't want to but did and found a broken basement window a few inches from the jar. As soon as the finger was warmed up the pain left.

The American psychologist William James carried out a survey of nearly 200 amputees in the 1880s, and found that beliefs of this kind were "very widespread."[30] More recently, some psychiatrists have attempted to explain phantom pain in terms of "fantasies" based on this belief. One case in the literature concerns a 14-year-old boy who suffered severe burning pain in the phantom following amputation of a leg. His psychiatrists discovered that in the previous year one of his schoolteachers had discussed amputation in class, and had told a story of a man in whom stinging pain had developed in a phantom limb. The leg was disinterred to find the cause of the pain, and ants were discovered burrowing into the amputated part. According to the story, the pain ceased when the ants were removed and the leg was carefully buried again. In the light of this story, the boy believed that the incineration of his amputated leg was the cause of the burning pain in the phantom.[31]

Another psychiatric case concerned a young woman who had had both legs amputated at the age of sixteen, following a car accident. She later suffered severe phantom pain, again of a burning kind. Under hypnosis, she recalled that at the time of the operation she had told the surgeon that she did not want her severed legs incinerated, but rather buried: she wanted to have a funeral for her legs. The surgeon ignored her request. The psychiatrist treated her by suggesting, under hypnosis, that in spite of her legs being incinerated, they were still with her in a spiritual sense, though not in the physical sense. "She reported increased feelings of well-being and seemed to believe that, symbolically, her legs had been restored to her." Her phantom pain disappeared completely.[32] This is one of the few cases of a complete cure that I have come across in the medical literature.

Similar beliefs are also widespread in present-day Russia, and probably in many other parts of the world. Of course, sceptics take it for granted that such beliefs are mere superstitions. But how can they be so sure? No one has ever done the appropriate experiments. Although it is not my principal purpose to explore the influence of severed parts on phantom pain, it is worth noting in passing that this question is in fact susceptible to empirical investigation.

Such experiments would not be hard to perform, given the cooperation of the staff and patients at a hospital where the severed limbs are routinely burned without consulting the patients. For the purposes of this experiment, the severed limbs would be divided at random into three groups. One lot of limbs would be burned as usual; the second would be buried straight; the third buried bent. This would be done following a "double-blind" procedure so that neither patients nor doctors knew the fate of the severed limbs. At various times thereafter, the patients would be questioned about their pain, if any. If there were no significant differences between groups, the sceptical hypothesis would be vindicated. But if there were differences such that those with burnt limbs had more burning pain, and those with limbs buried bent suffered more pain than those with limbs buried straight, then traditional folklore would receive experimental support. And medical practice could be altered accordingly, at least to the extent of offering patients some say in the way their severed limbs are disposed of.

PHANTOM LIMBS AND OUT-OF-THE-BODY EXPERIENCES

How are phantom limbs related to out-of-the-body experiences? In out-of-the-body experiences, people find themselves "outside" their bodies, implicitly or explicitly within a kind of phantom body.[33] Here, for example, is an account by a man who saw himself having an operation following a serious accident. He had become unconscious when given an anaesthetic. But total unconsciousness did not last long:

> I saw myself—my physical self—lying there. I saw a sharply outlined view of the operating table. I myself, freely hovering and looking downward from above, saw my physical body, lying on the operating table. I could see the wound of the operation on the right side of my body, see the doctor with an instrument in his hand, which I cannot more closely de-

scribe. All this I observed very clearly. I tried to hinder it all. It was so real. I can still hear the words I kept calling out: "Stop it—what are you doing there?"[34]

Some people can even "come out" of their physical bodies and move around at will. When the experience ends, they "re-enter" the physical body, and the phantom body fuses with it. One experienced out-of-the-body traveler is Robert Munroe,[35] who even gives practical courses on the subject at his center in Virginia, in the United States, training people how to travel out of their bodies. Here is his own description:

An out-of-body experience is a condition where you find yourself outside the physical body, fully conscious and able to perceive and act as if you were functioning physically with several exceptions. You can move through space (and time?) slowly or apparently somewhere beyond the speed of light. You can observe, participate in events, make willful decisions based upon what you perceive and do. You can move through physical matter such as walls, steel plates, concrete, earth, oceans, air, even atomic radiation without effort or effect. You can go into an adjoining room without bothering to open the door. You can visit a friend three thousand miles away. You can explore the moon, the solar system or the galaxy if these interest you.[36]

Out-of-the-body experiences are frequent in people who have nearly died, and are the starting point for so-called near-death experiences. This is how it felt to a 17-year-old boy who nearly drowned while swimming in a lake with some friends:

I kept bobbling up and down, and all of a sudden, it felt as though I were away from my body, away from everybody, in space by myself. Although I was stable, staying at the same level, I saw my body in the water about three or four feet away, bobbling up and down. I viewed my body from the

back and slightly from the right side. I still felt as though I
had an entire body form, even when I was outside my body.
I had an airy feeling that's almost indescribable. I felt like a
feather.[37]

Such experiences have been known in most, if not all, traditional
cultures. Even in modern industrial societies, they are far from
uncommon. Surveys have repeatedly shown that between 10 and
20 percent of the population remember having at least one out-of-
the-body experience.[38]

All of us have similar experiences in our dreams, when we seem
to travel far and wide even though our physical body is asleep in
bed. In our dreams we have a second body, the dream body. We
may not be aware of it all the time, just as we may not always be
aware of our physical body, but it is implicit. In our dreams we
have a location, a point of view, a center; we can move, see, hear,
talk. Sometimes we become particularly aware of the dream body,
as in flying or erotic dreams.

Some people have dreams in which they know that they are
dreaming, known as "lucid dreams." They still have a dream body,
but now they can will where they go, and to some extent control
their experience. Such dreams are very like out-of-the-body experiences, the main difference being that one is entered from the
dreaming state, and the other from the waking.[39]

In esoteric literature, traveling in lucid dreams or in out-of-the-
body experiences is known as "astral travel," and the body in
which this happens is called the "astral body" or "subtle body."
For many people this terminology is obscure and off-putting, and
in the following discussion I shall simply refer to the "non-material" body.

The similarities between the non-material body and phantom
limbs are striking. First, the phantom limb seems subjectively
real, and so does the non-material body, even though it is
known to be outside the flesh-and-blood body. Second, the
non-material body can separate from the normal body and then
re-enter it, just as in paraplegics and in patients with anaesthe-

tized nerves the phantom limb can separate from the normal limb and then re-enter it. And third, there are intermediate cases, especially when an injury to the spine has just happened: "Immediately after an accident, the phantom may be dissociated from the real body. For instance a person may feel as if the legs are raised over the chest or head even when he or she can see that they are stretched out on the road."[40]

The neurologist Ronald Melzack has concluded after many years of studying phantoms that: "It is evident that our experience of the body can occur without a body at all. We don't need a body to feel a body."[41] This is a matter of immediate experience for those who find themselves out of their bodies.

THEORIES OF PHANTOMS

What does all this mean? The answer depends very much on one's world view. For some people, the non-material body is an aspect of the psyche or soul. It normally animates the physical body but is capable of separating from it. Phantom limbs are aspects of the soul or psyche. They have a psychic rather than a material reality. This is probably the most widespread traditional view. Lord Nelson, the famous British admiral, lost an arm in a sea battle in 1797. He was fond of saying that, to him, his phantom arm was proof of the existence of the soul.

This interpretation of phantoms is still espoused by many psychics, some of whom claim to see the "auras" of missing limbs.[42] In esoteric circles, phantom limbs are regarded as aspects of the "subtle" or "astral" or "etheric" body.

By contrast, from the point of view of the contracted mind, phantoms and the non-material body are illusions generated within the nervous system. The phantoms are not where they seem to be: they are in the brain. For a committed materialist or mechanist, the brain theory is not so much a hypothesis as an article of faith: it *must* be true. Institutional medicine is still under the sway of the mechanistic theory, and hence the official view, taught to amputees by doctors and in hospitals, is that the phantom phenomenon is located inside the brain.

However, the exact location of phantoms has proved remarkably elusive. At first, the predominant hypothesis was that phantom limbs and phantom pain were caused by the generation of impulses by nerves in the remaining stump, particularly in nodules of nerves that grow at the cut ends, called neuromas. These impulses, flowing up through the spinal cord to the cerebral cortex, were supposed to generate sensations in the sensorimotor regions which were "referred" to the missing limb. This theory has repeatedly been tested, in attempts to relieve phantom pain, by surgically cutting the nerves from neuromas, either just above the neuroma or at the roots, next to the spinal cord. Although there is sometimes temporary relief, the phantoms persist and the pain usually returns. Moreover, the stump hypothesis cannot explain why some people born without limbs also experience phantoms in the absence of any injury to the nerves.

The next hypothesis moved the seat of the phantoms from the neuroma to the spinal cord, suggesting that the phantoms arose from spontaneous, excessive activity of nerves within the spinal cord that had lost their normal input from the body. Various nervous pathways within the spinal cord were cut in attempts to stop these effects, but the phantoms persisted and so did the pain. This hypothesis is also refuted by the experience of paraplegics whose spinal cord has been broken high up, for example in the neck. Some feel severe pain in the legs and groin, yet the spinal nerve cells that send impulses from those areas to the brain originate well below the level of the break, which means that any nerve impulses arising within them would not traverse the break.[43]

The hypothetical source of the phantoms has had to be shifted yet further back, into the brain. Areas of the thalamus and cerebral cortex that receive nerve impulses from the affected limb have been removed, but this ultimate surgical attempt to stop the pain has also failed. Even when the appropriate areas of the sensorimotor cortex are removed, the pain generally returns, and the phantom is still present![44]

Current versions of the brain theory push the supposed seat of the phantoms even further back, deeper into the cerebral tissues. One hypothesis proposes that the phantom depends on the way that new nerve connections are built up in the brain, "remapping" the areas which would previously have received nerve impulses from the amputated organ.[45] But the sprouting of new nerve connections would take weeks or months, and phantoms can appear immediately, as, for example, when the nerves serving a limb are anaesthetized. To avoid the need to invoke the sprouting of new nerves, another hypothesis proposes a rapid "unmasking of latent circuits" within extensive regions of the brain.[46] Yet another proposes that the body image is generated by a complex network of nerves in different parts of the brain, called a neuromatrix. The neuromatrix "generates patterns, processes information that flows through it, and ultimately produces the pattern that is felt as the whole body."[47] This neuromatrix is largely "hardwired." Although modified by experience, it is supposed to be innate, because people born without limbs can have phantoms of absent structures. It involves so much of the brain that to destroy the neuromatrix "would mean destruction of almost the whole brain."[48]

At this stage, the brain theory of phantoms becomes practically irrefutable. If the removal of any particular region of the brain fails to abolish the phantom, then it must be generated by other parts of the brain. "Parallel" or "back-up" or "latent" systems can be postulated indefinitely, rather as in pre-Copernican astronomy, epicycles could always be added to the supposed orbits of the planets to account for any awkward phenomena. Irrefutability is a virtue for committed believers, but a scientific vice.

In thinking about phantoms, medical researchers have been driven again and again to postulate concepts such as the "postural schema," "body schema," or "body image." The terms were introduced around the beginning of the century as a theoretical basis for explaining clinical observations, but their usage has remained extremely vague. In a critical review of the doctrine of the body schema, two eminent German neurologists have concluded:

There does not exist a well-defined and unitary theory of the body schema. On the contrary, various authors have developed quite different ideas resting upon quite different premises, intended to serve as explanation for different clinical phenomena. Moreover, the few really original contributions in this field have been subject to frequent misunderstandings and distortions. . . . Once this theory had been established, a great variety of disturbances were termed "disturbances of the body schema." These were then used to prove the validity of the theoretical concept. This is a classical case of *petitio principii,* in that one hypothesis served to explain another hypothesis and vice versa. Experimental investigations to test the theoretical hypotheses and their general validity without prejudice have been done very rarely.[49]

Freudians have their own particular interpretations of the body schema. This exists in "sensory-cerebral space-time" and involves "mental projection of the ego."[50] Phantoms are produced by the unconscious as a result of "a narcissistic desire to maintain the body's integrity in the face of a realistic loss or a rejection of symbolic castration of a body organ."[51] Such theories add to the terminology, but tell us next to nothing about the nature of the body image or the unconscious mind.

PHANTOMS AND FIELDS

All the conventional scientific theories are framed within the paradigm of the contracted mind: body schemas, images, and phantoms *must* be in the brain, irrespective of the most immediate experience. However, if the mind is extended within and beyond the body, there is no need to confine the body image to the brain or even to the nervous tissue. In particular, the phantom limb may not be confined to the brain but exist just where it seems to exist: projecting beyond the stump.

The extended mind resembles the traditional idea of the soul pervading and animating the body. But I think it is most helpful today to interpret this concept in terms of fields. The body is itself organized and pervaded by fields. As well as electromagnetic, gravitational, and quantum matter fields, morphogenetic fields shape its development and maintain its form. Behavioral, mental, and social fields underlie behavior and mental life. According to the hypothesis of formative causation, morphogenetic, behavioral, mental, and social fields are different kinds of morphic field, containing an inherent memory both from an individual's own past, and a collective memory from countless other people who have gone before.

Although I prefer to think of the fields of phantoms as morphic fields, the hypothesis I propose testing here is more general. I am not at present concerned with the specific feature of morphic fields, namely their habitual nature, shaped by morphic resonance. I am exploring the more general idea of fields as organizing patterns in space and time. I propose that these fields are located just where the phantoms seem to be. These fields can extend beyond the flesh-and-blood body, projecting beyond the stump.

A Simple Experiment
on the Effects of
Phantom Touch

The experiment I propose is analogous to that on the sense of being stared at, outlined in the previous chapter. Just as a person may be affected by being looked at, so a person may be affected by being "touched" by a phantom limb. Whatever the nature of the field that underlies the phantom, the person "touched" is organized by similar fields, and the fields of the amputee and the subject may interact.

The simplest form of this experiment is to follow the same general procedure as in the tests for the sense of being stared at. The subject sits behind a person with the phantom arm, and, in a random sequence, the person with the phantom either does nothing (control) or taps the subject on the shoulder with the phantom hand. The beginning of each trial is indicated by a click, buzz, or other mechanical signal. The subject then says whether he or she has felt the phantom touch or not. The result is recorded and the subject is told if the answer is right. This feedback should enable subjects to learn the unfamiliar feeling of a phantom touch—if it is possible to learn.

Of course, in the case of subjects with phantom legs rather than arms, the subject will be trying to detect a phantom touch from the foot, a phantom kick.

Results of a Preliminary Experiment

One of the amputees who wrote to me following my article in the *Bulletin of the Institute of Noetic Sciences* was Casimir Bernard of Hurley, New York. He lost his right leg below the knee on active service in Norway, as a member of the Allied Expeditionary Force in 1940. He had since worked as an expert in electronics manufacturing at IBM. He was already interested in psychical research, and was keen to try some experiments to find out if he could indeed touch someone with his phantom leg. He thought the experiment would best be conducted with a "sensitive" subject.

He discussed the matter with Dr. Alexander Imich of New York City, a retired chemist, who approached Ingo Swann, also living in New York City, who had taken part in a long series of apparently successful parapsychology experiments at the Stanford Research Institute in California. The three men worked together in designing and executing a series of tests, usually with Swann as the subject and Imich as the experimenter, but also with Imich as the subject and Swann as the experimenter. In these tests the subject attempted to feel Bernard's phantom leg. The experiments were carried out over several different days in March and April 1992.

The project has been written up by Swann as "An informal report of a preliminary experiment to sense a 'phantom limb.'" I am grateful to Ingo Swann, Alexander Imich, and Casimir Bernard for their permission to quote this report. Here is Swann's description of the procedure.

Mr. Casimir Bernard sat in a position in which he could raise or lower the phantom limb. The subject (Swann), with a hood covering his head to his shoulders, sat in a chair just in front of Mr. Bernard, in a position in which he could pass his right hand downward and back and forth through the limb if it was extended upward. The subject was then asked to call if his hand contacted the phantom limb. Dr. Imich silently signaled with a finger to Mr. Bernard either to raise or lower the limb. A bell was used to signal the subject to attempt each trial.

Rather than using a random number generator to determine whether in any given trial the phantom was to be raised or lowered, the experimenter made up a random-type sequence as he went along. The subject then called whether or not the limb was there. His calls were scored as correct or incorrect, and he could also pass, or in other words decline to answer. (Swann passed on 17 out of 175 trials, and Imich on 11 out of 96.) If he was correct, he was told. Thus there was the possibility of the subject learning to recognize the presence of the phantom as the experiment went along.

These are the gross average results, as given by Swann:

Swann: of 158 attempted calls, 89 were called correctly, when 79 would have been expected by chance.
Imich: of 84 attempted calls, 46 were called correctly, when 42 would have been expected by chance.

Swann also looked at the learning effect, which has often been found in psychic experiments at the Stanford Research Institute. Not surprisingly, psychic skills generally improve with learning, just as ordinary skills do. In his own words:

During my long tenure as an experimental designer at Stanford Research Institute, many characteristics signifying learning were studied and identified so that they could be reinforced. It was found that psychic learning progresses through subtle but predictable episodes, which appear to build on each other if they are reinforced with proper measures. Some of these indicators of learning are well–under-stood in general learning studies, but some are peculiar to psychic learning.

Swann plotted the cumulative number of correct calls on a graph, which also shows the line that would be expected on the basis of pure chance, with half the calls correct and half incorrect (Figure 11). With Swann as subject, a learning effect began to show up around trial 133. In the 25 trials from trial 133 until the end of the experiment, Swann was correct 22 times, against a chance expectation of 12.5. (I have examined the complete set of data statistically, taking the proportion of correct calls in successive groups of ten trials and analyzing the trend by means of linear regression. The tendency for Swann to be correct more often toward the end of the trial than toward the beginning, in other words the learning effect, is statistically significant with a probability of $p = 0.03$.)

With Imich as subject, the performance also improved with experience, with a learning effect occurring around trial 68. In the 17 trials from this point to the end of the experiment he was correct 11 times, against a chance expectation of 8.5.

As Swann points out: "If averaging of all trials is to be used to judge the success of this experiment, then it is not a notable success." But when the learning phenomenon is considered, especially with Swann himself as subject, the pattern of results "show that something was being learned and that this learning progressively enhanced the ability to determine whether the hands of the subject were interacting with the phantom limb." When the learning effect began to show up, Swann discovered that touching the phantom was unpleasant. He had no prior expectation whether it would be agreeable or not, but after this discovery, he found it easier to feel when the phantom was there, and his scores improved.

Of course sceptics will rightly want to know if the learning effect might have a more straightforward explanation. Could it simply be because the subject learned to use sound or other sensory cues? As Swann himself commented:

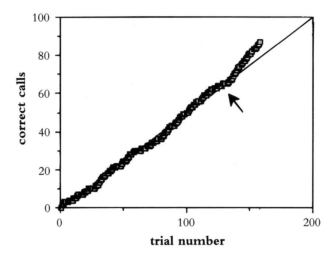

FIGURE XI

The cumulative number of times that Swann correctly identified whether Bernard's phantom leg was present or absent. Until around trial 133, he did no better than chance. But after this point, indicated by an arrow, when he said he had learned what the phantom felt like, his performance improved. The line indicates the proportion of correct guesses that would be expected on the basis of chance.

Visibility was completely prevented by the hood, but no feasible way was established to prevent hearing. On occasion Bernard's chair creaked, and the subject passed the call on each of these, for it was suggestive of the upward movement of the limb. Imich's room was overheated, and so the windows were open, allowing New York's street noises to be heard, and which masked room noise. But it would seem that the experiment was reasonably secure with regard to sensory cueing, for if not it would have been far easier to achieve a positive result earlier in the course of the trials.

But the possibility of subtle sensory cues could not be entirely excluded, nor could the possibility that the way the experimenter made up the sequence of trials, rather than using an independent method of randomization, introduced some subtle bias into the results.

Swann, Imich, and Bernard circulated this preliminary report to a number of researchers in parapsychology and medicine for their comments. The general consensus was that the experiment was interesting and the results encouraging, but that future experiments would need to use an independent method of randomization, eliminate possible sensory cues such as sound, and in some way control against the possibility that the effect was telepathic, involving the picking up of the amputee's intention to move the limb, or even the signal from the experimenter, rather than being due to the phantom itself. In addition, some researchers pointed out that there was no need for an experimenter at all. The amputee could be supplied with the random sequence directly, for example by means of a set of randomized cue-cards prepared in advance, and could also record the results.

I agree with these comments. My own suggestion for reducing the possibility of subtle cues is to do the experiment through a wall between two rooms, the more soundproof the better. If the phantom can still be detected when it is pushed through the wall, most kinds of sensory cues could be eliminated.

The sceptic within or without would then think of more arguments. Rather than a ghostly hand or foot emerging from the wall and actually being felt by the subject, there might still be some commonsense physical explanation. One obvious possibility: some sound signals might be getting through the wall. This could be tested most simply by asking subjects to wear earplugs. If sound is responsible, then the earplugs should reduce, if not abolish, the subject's apparent ability to feel the phantom. Another possibility: messages might be passed through some kind of mechanical vibration sensed with the whole body rather than with the ears. This could be tested by seating the subjects on layers of foam padding or some other vibration-damping material. And so on. Reasonable sceptical objections could be tested one by one, as long as the subjects retained enough interest and enthusiasm.

In order to test the possibility that subjects might pick up telepathically what the amputee is thinking about, rather than detecting the phantom itself, another experimental treatment could be included, so that the tests are done in three ways rather than two:

1. Control: phantom in rest position, as before. The amputee thinks of something else.
2. With the phantom extended, as before.
3. With the amputee thinking about moving the phantom, but not actually doing so. In addition the amputee could will the subject to feel the phantom.

Such experiments should reveal if there is indeed an effect of phantom touch over and above any possible effects of thinking or willing.

In my original suggestion for this experiment, I proposed that the subject be passive, trying to respond to a phantom touch from the amputee. However, the method used in the Bernard-Imich-Swann experiment involved active feeling for the phantom, and this might in general be a better method. Active feeling would be especially appropriate if the subjects were practitioners of "therapeutic touch" or other forms of subtle healing, who might be unusually sensitive to phantoms. Therapeutic touch is currently practiced by thousands of nurses and taught in many basic nursing programs in the United States. In response to my appeal for information, Dr. Barbara Joyce, head of the graduate nursing program at New Rochelle College, New York, has written to me about her experience with two women who had had their legs amputated. She tried to reduce pain and discomfort in their phantom legs.

In both instances patients reported that Therapeutic Touch used in the field of the missing limb reduces sensations of itching and pain. Although more clearly with one patient, but to some degree with the second, I was able to "feel" the phantom or missing limb and my estimation of its location in space corresponded with the patient's "sensation" of its location.

Perhaps not only experienced touch therapists, but people in general would do better if they were actively *feeling* for the phantom rather than passively waiting to be touched by it. So I suggest adopting an experimental design in which the subject feels for the phantom in a particular region of space and reports whether it is there or not. In preliminary tests, in which the subject tries to learn how to do this, the tests could be done in the same room, as in the Bernard-Imich-Swann experiment. But then, when the task is more familiar, the tests could be done through a wall, marking on the wall the place through which the phantom is to be pushed. In some trials, the phantom will be there; in others it will not; and in others the amputee will merely imagine it is there. The sequence of these trials will be determined by a standard randomizing procedure. The subject will then signal whether he or she feels the phantom or not, and will be informed when the call is correct.

Some Further
Experiments

1. If the phantom can indeed be felt after penetrating a barrier, then the effects of various kinds of barriers could be tested. Can the phantom pass through metal sheets? Can it pass through magnetized material? Can it pass through wires carrying electric currents? And so on.

2. If phantoms can be felt by other people, can the converse happen? Can an amputee feel when a person is "touching" or passing a hand through the phantom limb? Such an experiment could be subjected to controls similar to those already described.

3. Can the phantom be detected by various animals? In preliminary, informal tests, amputees could try touching their pets with their phantom limb. For example, if a sleeping cat or dog or horse is touched by a phantom hand, does it stir?

In this connection, I have heard from Mr. George Barcus of Toccoa, Georgia, that his constant companion, a little dog, "will not enter the area of my missing leg. She refuses to lay in the space vacated by the leg."

It might also be worth doing experiments with small animals that are particularly sensitive to human presence, reacting with alarm if a person comes close. Mice are one example, cockroaches another. If a phantom hand reaches into a cage of such animals through a barrier, do they show any kind of alarm reaction? Video recordings would be useful for a detailed analysis of any changes in their behavior.

4. Can the phantom be detected by physical means? For

example, does it affect the performance of sensitive pieces of physical apparatus? The simplest way to do initial tests would be to push the phantom into radios, televisions, computers, or other readily available machinery. Are there any observable effects? More sensitive tests could be done by putting the phantom in or near electrical and magnetic measuring instruments, or Geiger counters, mass spectrometers, nuclear magnetic resonance machines, bubble chambers used for detecting subatomic particles, etc. If the phantom interacts with the instrument, there should be different readings in the presence and absence of the phantom. And if such differences are found, the way will be open for increasingly sophisticated research on the phantom's physical properties.

5. Can the phantom be detected by Kirlian photography? This photographic technique employs high-voltage alternating current and depends on the recording of electrical discharges on film.[52] It is popular in New Age circles for "photographing the aura," and in many New Age festivals and psychic fairs you can have the "aura" of your hand photographed; the fee usually includes an interpretation of your emotional state. One of the most popular images in books and articles about Kirlian photography is the so-called phantom leaf. After a part of a leaf has been cut off, a phantom of the missing part still appears in the Kirlian image (Figure 12). This is a remarkable result, and would suggest that it might also be possible to photograph phantom limbs, fingers, etc.

There may be serious problems, though. The phantom leaf effect can arise by means of a simple error. If the operator first puts the leaf on the film and then cuts part of it off, this leaves a damp impression of the missing part. The image on the photograph simply arises from this dampness on the film.[53] Even pieces of damp blotting paper have an "aura" in a Kirlian photograph, and if the blotting paper is put down on the film first and then parts cut off, images of phantom blotting paper appear in photographs.

Although some "phantom leaf" images have been produced in this way, images of phantoms can still appear when the leaf is cut *before* being put down on the film. But not always. The effect is elusive, and while some practitioners can fairly frequently obtain phantom images, others do so rarely or not at all.[54] Several attempts have already been made to detect phantom limbs and fingers by this method, but so far without success.[55] So although the prospects for this line of research are not too hopeful, it might be worth a few more tries.

6. Can the phantom affect the germination of seeds or the growth of microorganisms? Phantom limbs could be pushed through trays of germinating seeds, or petri dishes containing bacterial cultures. Do the phantom-touched samples grow

FIGURE XII

A "phantom leaf." The upper part of the leaf was cut off along the line indicated by arrows, and then an image of the leaf was made by Kirlian photography. A ghostly image of the missing part appears to be present. (From a Kirlian photograph by Thelma Moss.)

significantly differently from untreated controls? Do bacterial cultures mutate more or less than controls untouched by the phantom? If so, do more frequent or longer exposures to the phantom have more effect than a single brief touch? And so on.

The Relationship of Mind and Body

The question underlying these experiments is what is the relationship of mind and body? Are our minds extended throughout our bodies or are they confined to the brain? Clearly, they *seem* to pervade our bodies. For example, if I feel pain in my big toe, I experience it in my toe, not in my brain. Likewise my general body awareness is experienced as being in my entire body, not merely inside my head. Nevertheless, the conventional view is that these subjective sensations arise within the brain, and are an aspect or epiphenomenon of brain processes.

Under normal circumstances there is not much that can be done to separate the experience of a limb from the physical limb itself. But this separation happens after amputations, or after the severing of nerves, or after some kinds of anaesthesia. The phantom limb can now be dissociated from the physical limb. Everyone agrees that this phantom limb has a subjective reality. But what underlies this? Does the experience of it arise only within the brain? Or is it associated with extended fields which pervade the body, and which continue to exist even when a structure is removed, just as the fields around magnets continue to exist even when the iron filings that reveal their presence are taken away?

The tests proposed in this chapter are designed to investigate whether or not the "subjective" phantom limb can have "objective" effects. If it can, phantoms will have to be regarded as more than mere processes in the brain, but as associated with fields located where the phantom seems to be.

The nature of these fields would be the next question. Are they extensions of known kinds of physical field, such as electromagnetic or quantum matter fields? Are they mental fields? Are they morphic fields, with an inherent memory? Are they all of these things together?

But obviously the principal question raised in this chapter needs to be answered first. Can phantoms have detectable effects? No one yet knows.

CONCLUSIONS

TO PART TWO

If people really can tell when they are being stared at and if phantom limbs can have detectable effects, then the paradigm of the contracted mind would have to be abandoned. The mind would reach out through the senses, projecting far beyond the surface of the body. It would pervade the body, in some sense animating it. The mind would no longer be shut up inside the brain. It would be liberated from its narrow confinement. The spell of Descartes would be broken.

The relation of mind, body, and environment would be seen in a new light. Vast new areas of medical, psychological, and philosophical research would open up. Parapsychology would find itself in a favorable scientific environment, rather than the usual hostile atmosphere. A great deal of folklore would need to be re-evaluated. A new understanding of the psyche would start to dawn. And the familiar separations between spirit and matter, mind and body, and subject and object would begin to break down.

On the other hand, the proposed experiments may fail. They

may not reveal the existence of any kinds of connection or communication at present unknown to physics. The sceptics may be vindicated. And so sceptics who believe in the importance of empirical inquiry should welcome these attempts to put their assumptions to the test.

SCIENTIFIC

ILLUSIONS

ILLUSIONS OF

OBJECTIVITY

PARADIGMS AND PREJUDICES

Many non-scientists are awed by the power and seeming certainty of scientific knowledge. So are most students of science. Textbooks are full of apparently hard facts and quantitative data. Science seems supremely objective. Moreover, a belief in the objectivity of science is a matter of faith for many modern people. It is fundamental to the worldview of materialists, rationalists, secular humanists, and all others who uphold the superiority of science over religion, traditional wisdom, and the arts.

This image of science is rarely discussed explicitly by scientists themselves. It tends to be absorbed implicitly and taken for granted. Few scientists show much interest in the philosophy, history, or sociology of science, and there is little room for these subjects in the crowded curriculum of science courses. Most simply assume that by means of "the scientific method," theories can be tested objectively by experiment in a way that is uncontaminated by the scientists' own hopes, ideas, and beliefs. Scientists like to think of themselves as engaged in a bold and fearless search for truth.

Such a view now excites much cynicism. But I think it is important to recognize the nobility of this ideal. Insofar as the scientific endeavor is illuminated by this heroic spirit, there is much to commend it. Nevertheless, in reality most scientists are now the servants of military and commercial interests.[1] Almost all are pursuing careers within institutions and professional organizations. The fear of career setbacks, rejection of papers by learned journals, loss of funding, and the ultimate sanction of dismissal are powerful disincentives to venture too far from current orthodoxy, at least in public. Many do not feel secure enough to voice their real opinions until they have retired, or won a Nobel Prize, or both.

Popular doubts about the objectivity of scientists are widely shared, for more sophisticated reasons, by philosophers, historians, and sociologists of science. Scientists are part of larger social, economic, and political systems; they constitute professional groups with their own initiation procedures, peer pressures, power structures, and systems of rewards. They generally work in the context of established paradigms or models of reality. And even within the limits set by the prevailing scientific belief system, they do not seek after pure facts for their own sake: they make guesses or hypotheses about the way things are, and then test them by experiment. Usually these experiments are motivated by a desire to support a favorite hypothesis, or to refute a rival one. What people do research on, and even what they find, is influenced by their conscious and unconscious expectations. In addition, feminist critics detect a strong and often unconscious male bias in the theory and practice of science.[2]

Many practicing scientists, like doctors, psychologists, anthropologists, sociologists, historians, and academics in general, are well aware that detached objectivity is more an ideal than a reflection of actual practice. In private, most are prepared to acknowledge that some of their colleagues, if not they themselves, are influenced in their researches by personal ambition, preconceptions, prejudices, and other sources of bias.

The tendency to find what is being looked for is deep-seated. It has a basis in the very nature of attention. The ability to focus the senses in accordance with intentions is a fundamental aspect of animal nature. Finding what is looked for is an essential feature of everyday human life. Most people are well aware that other people's attitudes affect the way they interact with the world around them. We are not surprised by such biases in politicians, nor by the differences in the way people see things within different cultures. We are not surprised to find many everyday examples of self-deception in members of our families and among friends and colleagues. But the "scientific method" is generally supposed to rise above cultural and personal biases, dealing only in the currency of objective facts and universal principles.

Biases in science are easiest to recognize when they reflect political prejudices, because people of opposing political views have a strong motive to dispute the claims of their opponents. For example, conservatives like to find a biological basis for the superiority of dominant classes and races, explaining their differences as largely innate. By contrast, liberals and socialists prefer to see environmental influences as predominant, explaining existing inequalities in terms of social and economic systems.

In the nineteenth century, this "nature–nurture" debate focused on measurements of brain size; in the twentieth, on measurements of IQ. Eminent scientists who were convinced of the innate superiority of men over women and of whites over other races, were able to find what they wanted to find. Paul Broca, for example, the anatomist after whom the speech area of the brain is named, concluded that: "In general, the brain is larger in mature adults than in the elderly, in men than in women, in eminent men than in men of mediocre talent, in superior races than in inferior races."[3] He had to overcome many factual obstacles to maintain this belief. For example, five eminent professors at Göttingen gave their consent to have their brains weighed after they died; when these cerebral weights turned out to be embarrassingly close to average, Broca concluded that the professors hadn't been so eminent after all!

Critics of a more egalitarian political persuasion have been able to show that generalizations based on different brain sizes or IQ scores rested on the systematic distortion and selection of data. Sometimes the data themselves were actually fraudulent, as in the case of some of the publications of Sir Cyril Burt, a leading defender of the view that intelligence is largely innate. In his book *The Mismeasure of Man,* Stephen Jay Gould traces the sorry history of these purportedly objective studies of human intelligence, showing how persistently prejudice has been clothed in scientific garb. "If—as I believe I have shown—quantitative data are as subject to cultural constraint as any other aspect of science, then they have no special claim on final truth."[4]

Public Pretense

A persistent and pervasive source of the illusion of objectivity is the style in which scientific reports are written. They give the appearance of coming from an idealized world in which science is an entirely logical exercise, free from all human passion. "Observations were made . . . ," "It was found that . . . ," "The data show . . . ," and so on. These literary conventions are still taught to budding scientists at school and university: "A test tube was taken. . . ."

Scientists publish their findings in technical articles, called papers, in specialized journals. In a justly famous essay called "Is the scientific paper a fraud?" the immunologist Peter Medawar pointed out that the standard structure of these papers gives "a totally misleading narrative of the processes of thought that go into the making of scientific discoveries." In the biological sciences, a typical paper starts with a brief introduction, including a survey of previous relevant work, then a section on "Materials and Methods," followed by "Results" and finally a "Discussion."

The section called "results" consists of a stream of factual information in which it is considered extremely bad form to discuss the significance of the results you are getting. You have to pretend that your mind is, so to speak, a virgin receptacle, an empty vessel, for information that floods in from the external world for no reason which you yourself have revealed. You reserve all appraisal of the scientific evidence until the "discussion" section, and in the discussion you adopt the ludicrous pretense of asking yourself if the information you have collected actually means anything.[5]

In fact, of course, the hypotheses that the experiments were designed to test generally come first, rather than last. Since Medawar wrote this passage there has been a greater conscious recognition of this sequence of events, and an increasing tendency to mention hypotheses in the introductions to papers. But the same conventions remain: passionless prose, the use of the passive voice, and the pretense that data are unembellished facts. Practicing scientists are well aware that this style is a kind of make-believe; but it has now become obligatory for anyone with pretensions to objectivity, and has been adopted by technocrats and bureaucrats as well.

DECEIT AND SELF-DECEPTION

The illusion of objectivity is most powerful when its victims believe themselves to be free of it. Along with a laudable sense of honor, a tendency to self-righteousness has been present in experimental science right from the outset.

With Galileo, the desire to make his ideas prevail apparently led him to report experiments that could not have been performed exactly as described. Thus an ambiguous attitude toward data was present from the very beginning of Western experimental science. On the one hand, experimental data was upheld as the ultimate arbiter of truth, on the other hand, fact was subordinated to theory when necessary and even, if it didn't fit, distorted.[6]

A similar vice afflicted other giants in the history of science, not least Sir Isaac Newton. He overwhelmed his critics with an exactness of results that left no room for dispute. His biographer Richard Westfall has documented how he adjusted his calculations on the velocity of sound and the precession of the equinoxes, and altered the correlation of a variable in his theory of gravitation to give a seeming accuracy of better than 1 part in 1,000.

Not the least part of the *Principia*'s persuasiveness was its deliberate pretense to a degree of precision quite beyond its legitimate claim. If the *Principia* established the quantitative pattern of modern science, it equally suggested a less sublime truth—that no one can manipulate the fudge factor so effectively as the master mathematician himself.[7]

Probably the commonest kind of deception—and of self-deception—depends on the selective use of data. For example, from 1910 to 1913, the American physicist Robert Millikan was engaged in a dispute with an Austrian rival, Felix Ehrenfeld, about the charge on the electron. Both Millikan's and Ehrenfeld's early data were rather variable. They depended on introducing oil drops into an electric field and measuring the strength of the field needed to keep them suspended. Ehrenfeld claimed that the data showed the existence of subelectrons with fractions of a unit electron charge. Millikan maintained there was a single charge. To rebut his rival, in 1913 he published a paper full of new, precise results supporting his own view, emphasizing in italics that "this is not a selected group of drops but represents all of the drops experimented upon during sixty consecutive days."[8]

A historian of science has recently examined Millikan's laboratory notebooks, which reveal a very different picture. The raw data were individually annotated with comments such as "very low, something wrong" and "beauty, publish this."[9] The 58 observations published in his article were selected from 140. Ehrenfeld meanwhile went on publishing all his observations, which continued to show a far greater variability than Millikan's selected data. Ehrenfeld was disregarded while Millikan won the Nobel Prize.

Millikan was no doubt convinced that he was right, and did not want his theoretical convictions to be disturbed by messy data. Probably the same was true of Gregor Mendel, the results of whose famous pea-breeding experiments were, according to modern statistical analysis, too good to be true.

The tendency to publish only the "best" results and to tidy up data is certainly not confined to famous figures in the history of science. In most if not all areas of science, good results are likely to advance the career of the person who produces them. And in a highly competitive and hierarchical professional environment, various forms of improving the results are widely practiced, if only by omitting unfavorable data. This practice is indeed normal. Apart from anything else, journals are disinclined to publish the results of problematical or negative experiments. Little professional credit results from unclear data or seemingly meaningless results.

I know of no formal study on the percentage of research data that are actually published. In the fields I know best from personal experience—biochemistry, developmental biology, plant physiology, and agriculture—I estimate that only about 5–20 percent of the empirical data are selected for publication. I have asked colleagues in other fields of inquiry, such as experimental psychology, chemistry, radioastronomy, and medicine, and come up with similar results. When the great majority of the data are discarded in private processes of selection—often 90 percent or more—there is obviously plenty of scope for personal bias and theoretical prejudice to operate both consciously and unconsciously.

The selective publication of data creates a context in which deception and self-deception become a matter of degree. Moreover, scientists usually regard their research notebooks and data files as private, and tend to resist any attempts by critics and rivals to go through them. True, it is usually assumed that a researcher will, within reason, make his or her data available to any colleague who might express a desire to see them. But in my own experience, this ideal is far from the reality. On the several occasions I have asked researchers if I may see their raw data, I have been refused. Maybe this says more about me than about prevailing scientific norms. But one of the very few systematic studies of this cherished principle of openness gives little ground for confidence. The procedure was simple. The person conducting it, a psychologist at Iowa State University, wrote to thirty-seven authors of papers published in psychology journals requesting the raw data on which the papers were based. Five did not reply. Twenty-one claimed that their data had unfortunately been misplaced or inadvertently destroyed. Two offered access only on very restrictive conditions. Only nine sent their raw data; and when their studies were analyzed, more than half had gross errors in the statistics alone.[10]

Those who refuse to expose their raw data to scrutiny may well have nothing to hide; they may simply find it inconvenient to explain their notes to someone else, or suspect the motives for the request, or resent an implied slur on their honor. It is not the purpose of this discussion to suggest that scientists are particularly prone to deliberate fraud and deception. On the contrary, most scientists are probably at least as honest as most members of other professional groups, such as lawyers, priests, bankers, and administrators. But scientists have greater pretensions to objectivity, and at the same time a culture which encourages the selective publication of results. These conditions are favorable for the deliberate deception of others, but I do not see this as the most important threat to the ideal of objectivity. Self-deception is the greatest danger, especially collective self-deception encouraged by dominant assumptions about objective reality.

Many scientists recognize the potential for wishful thinking in others and are quick to dismiss results of research in unorthodox fields such as parapsychology and holistic medicine as due to self-deception, if not deliberate fraud. And indeed some of those who challenge orthodox ideas may well deceive themselves. But they do little harm to the progress of science because their results are either ignored, or else subjected to extremely critical scrutiny. Organized groups of Skeptics, such as CSICOP, the Committee for the Scientific Investigation of Claims of the Paranormal, are always ready to challenge results that do not fit into the mechanistic worldview, and try their best to discredit them. Parapsychologists are so accustomed to these critical responses that they are unusually aware of the pitfalls of experimenter effects and other sources of bias. But conventional science is not subject to a similar degree of sceptical scrutiny.

PEER REVIEW, REPLICATION, AND FRAUD

Scientists, like doctors, lawyers, and other professionals, generally resist attempts by outside agencies to regulate their conduct. They pride themselves on their own system of controls. These are three-fold:

1. Applications for jobs and grants are subject to peer review, ensuring that the researchers and their projects meet with the approval of established professionals.
2. Papers submitted to scientific journals have to pass the critical scrutiny of expert referees, usually anonymous.
3. All published results are potentially subject to independent replication.

Peer review and refereeing procedures do indeed act as important quality checks, and are no doubt often effective, but they have a built-in bias. They tend to favor prestigious scientists and institutions. And independent replication is in fact rarely performed, for at least four reasons. First, in practice it is difficult to replicate a given experiment exactly, if only because the recipes are incomplete or fail to communicate practical knacks. Second, few researchers have the time or resources to repeat other people's work, especially if the results come from a well-funded laboratory and involve expensive apparatus. Third, there is usually no motivation for replicating the work of others. And fourth, even if exact replications are performed, it is difficult to get them published because scientific journals favor original research. Replication of other people's results is usually attempted only under special conditions, for example when the results are of unusual importance or when fraud is suspected on other grounds.

Under these circumstances deceptions can easily pass unchallenged as long as the results are in accordance with prevalent expectations.

> Acceptance of fraudulent results is the other side of that familiar coin, resistance to new ideas. Fraudulent results are likely to be accepted in science if they are plausibly presented, if they conform with prevailing prejudices and expectations, and if they come from a suitably qualified scientist affiliated with an elite institution. It is for the lack of all these qualities that new ideas in science are likely to be resisted. Only on the assumption that logic and objectivity are the sole gatekeepers of science is the prevalence and frequent success of fraud in any way surprising. . . . For the ideologists of science, fraud is taboo, a scandal whose significance must be ritually denied on every occasion. For those who see science as a human endeavor to make sense of the world, fraud is merely evidence that science flies on the wings of rhetoric as well as reason.[11]

One of the few areas of science subject to a limited form of external supervision is the testing for safety of new foods, drugs, and pesticides. In the United States, every year many thousands of test results are submitted by industry for review by the Food and Drug Administration (FDA) or the Environmental Protection Agency (EPA). These agencies have the power to send inspectors to the laboratories that provide the data. They continually unearth falsified results.[12]

The cases of fraud uncovered in the great unpoliced hinterlands of science are rarely brought to light by the official mechanisms of peer review, refereeing of papers, and the potential for independent replication. And even if attempts to replicate an experiment fail, this is usually ascribed to a failure to reproduce the conditions of the experiment precisely enough. There is a big psychological and cultural barrier against accusing colleagues of fraud—unless one has strong personal reasons to suspect their integrity. Most known cases of fraud come to light as a result of whistle-blowing by immediate colleagues or rivals, often as a result of some personal grievance.[13] When this happens, the typical response of laboratory chiefs and other responsible authorities is to try to hush the matter up. But if the charges of fraud do not blow over, if allegations are made persistently enough, and if the evidence becomes over-whelming, then an official inquiry is held. Someone is found guilty and dismissed in disgrace.

Most professional scientists deny that these incidents shed doubt on institutional science as a whole; rather, they are seen as isolated aberrations by individuals who have become temporarily unhinged under pressure, or who are rare but inevitable psychopaths. Science is purified by their expulsion. They are scapegoats in the biblical sense. On the Day of Atonement the high priest confessed the sins of the people while laying his hands on a goat. The guilt-laden scapegoat was then expelled from the community into the wilder-ness, bearing away their iniquities.[14]

Scientists generally feel the need to preserve an idealized self-image, not just for personal and professional reasons, but also be-cause this image is projected on to them by others. There are many people who put their faith in science rather than religion, and need to believe in its superior, objective authority. And to the extent that science replaces religion as the source of truth and values, then scientists become a kind of priesthood. As with priests in general, there is then a public expectation that they will live up to the ideals they preach: in the case of scientists, objectivity, rationality, and the quest for truth. "Some scientists, in their public appearances, can be noticed playing up to this role, which seems to invest them as cardinals of reason propounding salvation to an irrational pub-lic."[15] There is also a strong disincentive for them to admit that there is anything fundamentally wrong with the beliefs and institu-tions that legitimize their own position. While it is relatively easy to admit that individuals may err, and to purify the community by expelling them, it is much harder to question the beliefs and idealizations on which the whole system depends.

Philosophers of science tend to idealize the experimental method, and so do scientists themselves. In their insightful study of fraud and deceit in science, William Broad and Nicholas Wade were led to inquire what actually happens in laboratories, as opposed to what is supposed to happen. They found that the reality was far more pragmatic and empirical, involving much trial and error:

> The competitors in a given field try many different approaches but are always quick to switch to the recipe that works best. Science being a social process, each researcher is trying at the same time to advance and gain acceptance for his own recipes, his own interpretation of the field. . . . Science is a complex process in which the observer can see almost anything he wants provided he narrows his vision sufficiently. . . . Scientists are individuals and they have different styles and different approaches to the truth. The identical style of all scientific writing, which seems to spring from a universal scientific method, is a false unanimity imposed by the current conventions of scientific reporting. If scientists were allowed to express themselves naturally in describing their experiments and theories, the myth of a single, universal scientific method would probably vanish instantly.[16]

I agree with this analysis. This present book is an argument for more democratic and pluralistic scientific research, liberated from the conventions imposed upon institutional science by its role as the Established Church of the secular world order. But whatever forms science takes, it will still depend on experiments.

EXPERIMENTS ON EXPERIMENTS

So far in this discussion I have considered the general problems caused by the illusion of objectivity. In the following two chapters I outline experiments to investigate the nature of experimental research itself.

In chapter 6, I consider the doctrine of uniformity, which biases scientists against seeing unexpected patterns or irregularities in nature. Even the constancy of the "fundamental constants" turns out to be a matter of faith. These constants, as actually measured, fluctuate. Treating variations as random errors enables the data to be smoothed out, concealing underlying variations behind a uniform façade. I suggest a way in which these observed variations can be investigated empirically.

In chapter 7, I turn to the influence of the expectations of experimenters on experiments themselves. These may well include subtle influences, perhaps including paranormal effects, on the system under study. How much do experiments tell us about nature, and how much do they merely reflect the expectations of the experimenter?

CHAPTER SIX

THE VARIABILITY

OF THE "FUNDAMENTAL

CONSTANTS"

THE FUNDAMENTAL PHYSICAL CONSTANTS
AND THEIR MEASUREMENT

The "physical constants" are numbers used by scientists in their calculations. Unlike the constants of mathematics, such as π, the values of the constants of nature cannot be calculated from first principles; they depend on laboratory measurements.

As the name implies, the so-called physical constants are supposed to be changeless. They are believed to reflect an underlying constancy of nature. In this chapter I discuss how the values of the fundamental physical constants have in fact changed over the last few decades, and suggest how the nature of these changes can be investigated further.

There are many constants listed in handbooks of physics and chemistry, such as melting points and boiling points of thousands of chemicals, going on for hundreds of pages: for instance the boiling point of ethyl alcohol is 78.5°C at standard temperature and pressure; its freezing point is −117.3°C. But some constants are more fundamental than others. The following list gives the seven most generally regarded as truly fundamental (Table 1).[1]

All these constants are expressed in terms of units; for example, the velocity of light is expressed in terms of meters per second. If the units change, so will the constants. And units are man-made, dependent on definitions that may change from time to time: the meter, for instance, was originally defined in 1790 by a decree of the French National Assembly as one ten-millionth of the quadrant of the earth's meridian passing through Paris. The entire metric system was based upon the meter and imposed by law. But the original measurements of the earth's circumference were found to be in error. The meter was then defined, in 1799, in terms of a standard bar kept in France under official supervision. In 1960 the meter was redefined in terms of the wavelength of light emitted by krypton atoms; and in 1983 it was redefined again in terms of the speed of light itself, as the length of the path traveled by light in 1/299,792,458 of a second.

As well as any changes due to changing units, the official values of the fundamental constants vary from time to time as new mea-

TABLE ONE

THE FUNDAMENTAL CONSTANTS

Fundamental quantity	Symbol
Velocity of light	c
Elementary charge	e
Mass of the electron	m_e
Mass of the proton	m_p
Avogadro constant	N_A
Planck's constant	h
Universal gravitational constant	G
Boltzmann's constant	k

surements are made. They are continually adjusted by experts and international commissions. Old values are replaced by new ones, based on the latest "best values" obtained in laboratories around the world. Below, I consider in detail four examples: the gravitational constant *(G);* the speed of light *(c);* Planck's constant *(h);* and also the fine structure constant (α), which is derived from the charge on the electron *(e),* the velocity of light, and Planck's constant.

The "best" values are already the result of considerable selection. First, experimenters tend to reject unexpected data on the grounds that they must be errors. Second, after the most deviant measurements have been weeded out, variations within a given laboratory are smoothed out by averaging the values obtained at different times, and the final value is then subjected to a series of somewhat arbitrary corrections. Finally, the results from different laboratories around the world are selected, adjusted, and averaged to arrive at the latest official value.

The measurement of fundamental constants is the province of specialists known as metrologists. In the past, the field was dominated by individuals, such as the American R. T. Birge of the University of California at Berkeley, who reigned supreme from the 1920s to the 1940s. Nowadays the final values are set by international committees of experts. At any given time, the official values of the constants depend on a series of decisions by experimenters themselves, by the doyens of metrology, and by committees. As Birge described the process:

> The decision as to the most probable value, at a particular time, of any given constant, necessarily demands a certain amount of judgement. . . . Similarly, each investigator uses a certain amount of judgement in the selection of his data and in the final conclusions reached.[2]

Faith in Eternal Truths

In practice, then, the values of the constants change. But in theory they are supposed to be changeless. The conflict between theory and empirical reality is usually brushed aside without discussion, because all variations are assumed to be due to experimental errors, and the latest values are assumed to be the best. Metrologists are treated with limitless indulgence. Past values of the constants are soon forgiven and forgotten.

But what if the constants *really* change? What if the underlying nature of nature varies? Before this subject can even be discussed, it is necessary to think about one of the most fundamental assumptions of science as we know it: faith in the uniformity of nature. For the committed believer, these questions are nonsensical. Constants *must* be constant.

Most constants have been measured only in this small region of the universe for a few decades, and the actual measurements have varied erratically. The idea that all constants are the same everywhere and always is not an extrapolation from the data. If it *were* an extrapolation it would be outrageous. The values of the constants as actually measured on earth have changed considerably over the last fifty years. To assume they had not changed for fifteen billion years anywhere in the universe goes far beyond the meager evidence. The fact that this assumption is so little questioned, so readily taken for granted, shows the strength of scientific faith in eternal truths.

According to the traditional creed of science, everything is governed by fixed laws and eternal constants. The laws of nature are the same in all times and at all places. In fact they transcend space and time. They are more like eternal Ideas—in the sense of Platonic philosophy—than evolving things. They are not made of matter, energy, fields, space, or time; they are not made of anything. In short, they are immaterial and non-physical. Like Platonic Ideas they underlie all phenomena as their hidden reason or *logos,* transcending space and time.

Of course, everyone agrees that the laws of nature as formulated by scientists change from time to time, as old theories are partially or completely superseded by new ones. For example, Newton's theory of gravitation, depending on forces acting at a distance in absolute time and space, was replaced by Einstein's theory of the gravitational field consisting of curvatures of space-time itself. But both Newton and Einstein shared the Platonic faith that underlying the changing theories of natural science there are true eternal laws, universal and immutable. And neither challenged the constancy of constants: indeed both gave great prestige to this assumption, Newton through his introduction of the universal gravitational constant, and Einstein through treating the speed of light as an absolute. In modern relativity theory, c is a mathematical constant, a parameter relating the units used for time to the units used for space; its value is fixed by definition. The question as to whether the speed of light actually differs from $c,$ although theoretically conceivable, seems of peripheral interest.

For the founding fathers of modern science, such as Copernicus, Kepler, Galileo, Descartes, and Newton, the laws of nature were changeless Ideas in the divine mind. God was a mathematician. The discovery of the mathematical laws of nature was a direct insight into the eternal Mind of God.[3] Similar sentiments have been echoed by physicists ever since.[4]

By the end of the eighteenth century, many intellectuals subscribed to a belief known as Deism, with a remote, rational, mathematical deity stripped of the troublesome attributes of the biblical God. This supreme being was knowable through human reason without the need for divine revelation or religious institutions. The God of Deism created the universe in the first place, but thereafter played no active role. Everything took place automatically in accordance with the laws and constants of nature. These laws, as aspects of the divine mind, participated in the divine attributes; they were absolute, universal, changeless, and omnipotent.

From the early nineteenth century onwards, Deism increasingly gave way to atheism. God became an "unnecessary hypothesis," as the French physicist Henri Laplace put it. The eternity of matter and energy was guaranteed by the principles of conservation of matter and energy; the eternity of the laws of nature and the constancy of the constants were simply taken for granted. The immaterial mathematical principles of nature were somehow free-floating, self-sustaining, and mysteriously mind-like—and potentially knowable by mathematicians.

Until the 1960s, the universe of orthodox physics was still eternal. But evidence for the expansion of the universe has been accumulating for several decades, and the discovery of the cosmic microwave background radiation in 1965 finally triggered off a great cosmological revolution. The Big Bang theory took over. Instead of an eternal machine-like universe, gradually running down toward a thermodynamic heat death, the picture was now one of a growing, developing, evolutionary cosmos. And if there was a birth of the cosmos, an initial "singularity," as physicists put it, then once again age-old questions arise. Where and what did everything come from? Why is the universe as it is? In addition, a new question arises. If all nature evolves, why should the laws of nature not evolve as well? If laws are immanent in evolving nature, then the laws should evolve too.

For the most part, physicists still take the traditional Platonic approach. The laws do not arise out of the evolving cosmos but are imposed upon it. They were there to start with, like a kind of cosmic Napoleonic Code. Somehow, out of an eternal, non-physical mind-like realm—the mind of a mathematical God, or just a free-floating realm of mathematics—the universe was born from a void in a primal explosion. This is how the physicist Heinz Pagels expressed it:

> The nothingness "before" the creation of the universe is the most complete void that we can imagine—no space or time or matter existed. It would be a world without place, without duration or eternity, without number—it is what mathemati-

cians call "the empty set." Yet this unthinkable void converts itself into a plenum of existence—a necessary consequence of physical laws. Where are these laws written into that void? What "tells" the void that it is pregnant with a possible universe? It would seem that even the void is subject to law, a logic that exists prior to time and space.[5]

Modern attempts to create a mathematical Theory of Everything accept an evolutionary cosmology, but at the same time accept the traditional faith in eternal laws of nature and the invariance of the fundamental constants. The laws were in some sense already there before the initial singularity; or rather they transcend time and space altogether. But the question remains: why should the laws be as they are? And why should the fundamental constants have the particular values they have?

Today these questions are usually discussed in terms of the anthropic cosmological principle, as follows: Out of the many possible universes, only one with the constants set at the values found today could have given rise to a world with life as we know it and allowed the emergence of intelligent cosmologists capable of discussing it. If the values of the constants had been different, there would have been no stars, nor atoms, nor planets, nor people. Even if the constants were only slightly different, we would not be here. For example, with just a small change in the relative strengths of the nuclear and electromagnetic forces there could be no carbon atoms, and hence no carbon-based forms of life such as ourselves. "The Holy Grail of modern physics is to explain why these numerical constants . . . have the particular numerical values they do."[6]

Some physicists incline toward a kind of neo-Deism, with a mathematical creator-God who fine-tuned the constants in the first place, selecting from many possible universes the one in which we can evolve. Others prefer to leave God out of it. One way of avoiding the need for a mathematical mind to fix the constants of nature is to suppose that our universe arose from a froth of possible universes. The primordial bubble that gave rise to our universe was one of many. But our universe has to have the constants it does by the very fact we are here. Somehow our presence imposes a selection. There may be innumerable alien and uninhabitable universes quite unknown to us, but this is the only one we can know.

This kind of speculation has been carried even further by Lee Smolin, who has proposed a kind of cosmic Darwinism. Through black holes, baby universes may be budded off from pre-existing ones and take on a life of their own. Some of these might have slight mutations in the values of their constants and hence evolve differently. Only those that form stars can form black holes and hence have babies. So by a principle of cosmic fecundity, only universes like ours would reproduce, and there may be many more or less similar habitable universes.[7] But this very speculative theory still does not explain why any universes should exist in the first place, nor what determines the laws that govern them, nor what maintains, carries, or remembers the mutant constants in any particular universe.

Notice that all these metaphysical speculations, extravagant though they seem, are thoroughly conventional in that they take for granted both eternal laws and constant constants, at least within a given universe. These well-established assumptions make the constancy of constants seem like an assured truth. Their changelessness is an act of faith, grounded in Platonic philosophy and theology. But this belief goes far beyond the evidence. Even in the last few decades, the official values of the constants have changed. And attempts to measure constants over astronomical distances and times, using astronomical methods, are all based on the assumption that they set out to prove, namely a universal constancy of nature. They are, to varying degrees, based on circular arguments, as I show below. But mere empirical data have little to do with the faith of the committed believer. If measurements show variations in the constants, as they often do, then the variations are dismissed as experimental errors; the latest figure is the best available approximation to the "true" value of the constant.

Some variations may well be due to errors, and such errors decrease as instruments and methods of measurement improve. All kinds of measurements have inherent limitations on their accuracy. But not all the variations in the measured values of the constants need necessarily be due to error, or to the limitations of the apparatus used. Some may be real. In an evolving universe, it is conceivable that the constants evolve along with nature. They might even vary cyclically, if not chaotically.

THEORIES OF CHANGING CONSTANTS

Several physicists, among them Arthur Eddington and Paul Dirac, have speculated that at least some of the "fundamental constants" may change with time. In particular, Dirac proposed that the universal gravitational constant, G, may be decreasing with time: the gravitational force weakening as the universe expands.[8] But those who make such speculations are usually quick to avow that they are not challenging the idea of eternal laws; they are merely proposing that eternal laws govern the variation of the constants.

The proposal that the *laws* themselves evolve is more radical. The philosopher Alfred North Whitehead pointed out that if we drop the old idea of Platonic laws imposed on nature, and think instead of laws being immanent in nature, then they *must* evolve along with the nature:

> Since the laws of nature depend on the individual characters of the things constituting nature, as the things change, then consequently the laws will change. Thus the modern evolutionary view of the physical universe should conceive of the laws of nature as evolving concurrently with the things constituting the environment. Thus the conception of the Universe as evolving subject to fixed eternal laws should be abandoned.[9]

I prefer to drop the metaphor of "law" altogether, with its outmoded image of God as a kind of law-giving emperor, as well as an omnipotent and universal law-enforcement agency. Instead, I have suggested that the regularities of nature may be more like habits. According to the hypothesis of morphic resonance, a kind of cumulative memory is inherent in nature. Rather than being governed by an eternal mathematical mind, nature is shaped by habits, subject to natural selection.[10] And some habits are more fundamental than others; for example, the habits of hydrogen atoms are very ancient and widespread, found throughout the universe, while the habits of hyenas are not. Gravitational and electromagnetic fields, atoms, galaxies, and stars are governed by archaic habits, dating back to the earliest periods in the history of the universe. From this point of view the "fundamental constants" are quantitative aspects of deep-seated habits. They may have changed at first, but as they became increasingly fixed through repetition, the constants may have settled down to more or less stable values. In this respect the habit hypothesis agrees with the conventional assumption of constancy, though for very different reasons.

Even if speculations about the evolution of constants are set aside, there are at least two more reasons why constants may vary. First, they may depend on the astronomical environment, changing as the solar system moves within the galaxy, or as our galaxy moves away from other galaxies. And second, the constants may oscillate or fluctuate. They may even fluctuate in a seemingly chaotic manner. Modern chaos theory has enabled us to recognize that chaotic behavior, as opposed to old-style determinism, is normal in most realms of nature.[11] So far the "constants" have survived unchallenged from an earlier era of physics: the vestiges of a lingering Platonism. But what if they, too, vary chaotically?

The possibility that constants change slightly over many millions of years has been taken seriously by metrologists, and various attempts have been made to estimate possible changes indirectly, for example by comparing the light from stars and galaxies believed to be relatively near and far away, and hence differing in age by many millions or even billions of light years. Such methods suggest that if there are systematic changes in the fundamental constants, they are very small. But the trouble is that these indirect methods depend on many assumptions that cannot be checked directly. To varying degrees, the indirect evidence for the constancy of constants depends on circular arguments. I look at this evidence in more detail in my discussion of particular constants.

Even if the average values of the constants are in fact stable over long periods of time, the values at any given time may vary around these averages as a result of changing extraterrestrial environments, or fluctuate chaotically. But what about the facts?

The Variability of the Universal Gravitational Constant

This universal gravitational constant, G, first appeared in Newton's gravitation equation, which states that the force of gravitational attraction is equal to G times the product of the attracting masses divided by the square of the distance separating them. The value of this constant has been measured many times since the first precision experiment of Henry Cavendish in 1798. The "best" values over the last century or so are shown in Figure 13. There was initially a wide spread of values, and then a convergence toward a consensus figure. But even since 1970 the "best" values have ranged from 6.669 9 to 6.674 5, a difference of 0.07 per cent.[12] (The units in which these values are expressed are $\times 10^{-11}$ m^3 kg^{-1} s^{-2}.)

In spite of the central importance of the universal gravitational constant, it is the least well defined of all the fundamental constants. Attempts to pin it down to many places of decimals have failed; the measurements are just too variable. The editor of the scientific journal *Nature* has described as "a blot on the face of physics" the fact that G still remains uncertain to about one part in 5,000.[13] Indeed, in recent years the uncertainty has been so great that the existence of entirely new forces has been postulated to explain gravitational anomalies.

In the early 1980s, Frank Stacey and his colleagues measured G in deep mines and boreholes in Australia. Their value was about 1 percent higher than that currently accepted. For example, in one

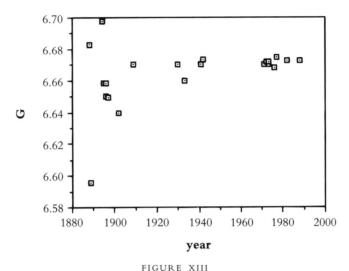

FIGURE XIII

*Best values of the universal gravitational
constant (G) from 1888 to 1989.*

set of measurements in the Hilton mine in Queensland the value of
G was found to be 6.734 ± 0.002, as opposed to the currently
accepted value of 6.672 ± 0.003.[14] The Australian results were
repeatable and consistent,[15] but no one took much notice until
1986. In that year Ephrain Fischbach, at the University of Washington, Seattle, sent shock waves around the world of science by
claiming that laboratory tests also showed a slight deviation from
Newton's law of gravity, consistent with the Australian results. He
and his colleagues reanalyzed the data from a series of experiments
by Roland Eötvös in the 1920s, one of the standard textbook
examples of exact measurement, and found that there was a consistent anomaly hidden in the data that had been dismissed as random
error.[16] On the basis of these laboratory results and observations
from Australian mines, Fischbach proposed the existence of a hitherto unknown repulsive force, the so-called fifth force (the four
known forces being the strong and weak nuclear forces, the electromagnetic force, and the gravitational force).

In the following years, further careful measurements of gravitation in deep mines, in holes in the Arctic ice cap, and on tall
towers provided more evidence for the fifth force.[17] The interpre-

tation of these results depended on taking into account the local geology, since the density of the surrounding rocks affects the measurements of gravitation. The experimenters were well aware of this, and carried out appropriate corrections. But sceptics argued that the results were due to hidden rocks of unusually high density, and postulated buried outcrops to explain the results.[18] For the time being, the sceptical view predominates, although the existence of the fifth force is still an open question and the subject of considerable theoretical and experimental research.[19]

The possible existence of a fifth force is not particularly relevant to possible changes in G with time. But the very fact that the question of an extra force affecting gravitation could even be raised and seriously considered in the late twentieth century serves to emphasize how imprecise the characterization of gravity remains more than three centuries after the publication of Newton's *Principia*.

The suggestion by Paul Dirac and other theoretical physicists that G may be decreasing as the universe expands has been taken quite seriously by some metrologists. However, the change proposed by Dirac was very small, about 5 parts in 10^{11} per year. This is way below the limits of detection using conventional methods of measuring G on Earth. The "best" results in the last twenty years differ from each other by more than 5 parts in 10^4. In other words, the change Dirac was suggesting is some ten million times smaller than the differences between recent "best" values.

In order to test Dirac's hypothesis, a variety of indirect methods have been tried. Some depend on geological evidence, such as the slopes of fossil sand dunes, from which the gravitational forces at the time they were formed can be calculated; others depend on records of eclipses over the last 3,000 years; others on modern astronomical methods. In one, the distance of the moon was monitored at regular intervals using a sophisticated form of radar made possible by placing an array of reflectors on the moon's surface as part of the space program. The time of flight of laser pulses, launched and detected by a telescope, were measured at regular intervals. A more accurate radar technique resulted from the *Viking* mission to Mars, with pulses transmitted back to the earth from landers on the surface of that planet. Such measurements were continued from 1976 to 1982. By assuming a fixed value for the speed of light, these radar techniques enabled the distance from Mars to the earth to be monitored to a precision of several meters. Then, on the basis of complex mathematical models of the orbits of the various bodies in the solar system, the data were checked to see if they were consistent with a constant value for G. But the calculations involved many uncertainties, including assumptions about the interference with the orbit of Mars by large asteroids of unknown mass. One way of calculating the data gave results consistent with G varying by less than 0.2 parts in 10^{11} per year.[20] Another calculation using the same data pointed to a variation more than ten times greater, but still less than 1 part in 10^{10} per year.[21]

Yet another astronomical method is to study the dynamics of a distant binary pulsar to see if these are consistent with a constant value of G over the period of observation. Again a great many assumptions are necessary in order to perform the calculations, which makes them questionable by anyone who wants to change the assumptions.[22]

Some physicists think that at least some of the data point to small changes in G with time.[23] On the basis of lunar data, some have concluded that G may be changing at least as much as Dirac proposed;[24] others think it is not.[25] These various studies have been interpreted by the doyen of British metrology, Brian Petley, as follows:

> Providing the cosmological time scales can be relied on and that our understanding of gravitation is sufficient, the variations of G are less than around 1 part in 10^{10} per year. This conclusion is supported by a range of different evidence, some of it from quite short-term experiments. If the change postulated by Dirac is ruled out, one is left with changes in G to some small power of time, or perhaps to postulate a cyclic variation with little variation at the present epoch.[26]

The problem with all these indirect lines of evidence is that they depend on a complex tissue of theoretical assumptions, including the constancy of the other constants of nature. They are persuasive only within the framework of the present paradigm. That is to say that if one assumes the correctness of modern cosmological theories, themselves presupposing the constancy of G, the data are internally consistent, provided that all actual variations from experiment to experiment, or method to method, are assumed to be a result of error.

THE FALL IN THE SPEED OF LIGHT
FROM 1928 TO 1945

According to Einstein's theory of relativity, the speed of light in a vacuum is invariant: it is an absolute constant. Much of modern physics is based on this assumption. There is therefore a strong theoretical prejudice against raising the question of possible changes in the velocity of light. In any case, the question is now officially closed. Since 1972 the speed of light has been fixed *by definition*. The value is defined as 299,792.458 ± 0.001 # 2 kilometers per second.

As in the case of the universal gravitational constant, early measurements of c differed considerably from the present official value. For example, the determination by Römer in 1676 was about 30 percent lower, and that by Fizeau in 1849 about 5 percent higher.[27] The progress of the "best" values since 1874 is shown in Figure 14. At first sight this seems to be another brilliant example of the progress of exact science, coming closer and closer to the truth. But the detailed facts are more complex.

In 1929, Birge published his review of all the evidence available up to 1927 and came to the conclusion that the best value for velocity of light was 299,796 ± 4 km/s. He pointed out that the probable error was far less than in any of the other constants, and concluded that "the present value of c is entirely satisfactory, and can be considered as more or less permanently established."[28]

However, even as he was writing, considerably lower values of c were being found, and by 1934 it was suggested by Gheury de Bray that the data pointed to a cyclic variation in the velocity of light.[29]

From around 1928 to 1945, the velocity of light appeared to be about 20 km/s lower than before and after this period (Table 2). The "best" values, found by the leading investigators using a variety of techniques, were in impressively close agreement with each other, and the available data were combined and adjusted by Birge in 1941 and Dorsey in 1945.

In the late 1940s the speed of light went up again. Not surprisingly, there was some turbulence at first as the old value was overthrown. The new value was about 20 km/s higher, close to that prevailing in 1927. A new consensus developed (Figure 15). How long this consensus would have lasted if based on continuing measurements is a matter for speculation. In practice, further disagreement was prevented by fixing the speed of light in 1972 by definition.

How can the lower velocity from 1928 to 1945 be explained? If it was simply a matter of experimental error, why did the results of

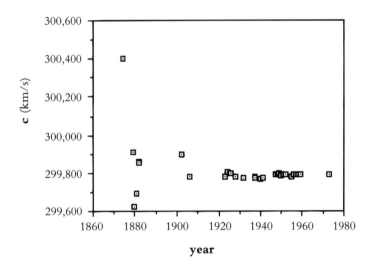

FIGURE XIV

Best values of the velocity of light (c) from 1874 to 1972.

TABLE TWO

THE SPEED OF LIGHT, 1928–1945[30]

Author	Date	Velocity of light (km/s)
Previous accepted value (Birge, 1929)		299,796 4
Mittelstaedt	1928	299,778 ± 20
Michelson *et al.*	1932	299,774 ± 11
Michelson *et al.*	1935	299,774 ± 4
Anderson	1937	299,771 ± 10
Hüttel	1940	299,771 ± 10
Anderson	1941	299,776 ± 6
Birge (review)	1941	299,776 ± 4
Dorsey (review)	1945	299,773 ± 10
Present defined value, 1972 onwards		299,792.458 ± 0.001 2

different investigators and different methods agree so well? And why were the estimated errors so low?

One possibility is that the velocity of light really does fluctuate from time to time. Perhaps it really did drop for nearly twenty years. But this is not a possibility that has been seriously considered by researchers in the field, except for de Bray. So strong is the assumption that it must be fixed that the empirical data have to be explained away. This remarkable episode in the history of the speed of light is now generally attributed to the psychology of metrologists:

The tendency for experiments in a given epoch to agree with one another has been described by the delicate phrase "intel-

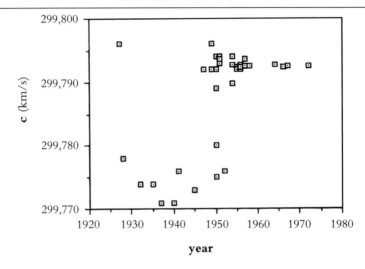

FIGURE XV

Values of the velocity of light (c) from 1927 to 1972. In 1972 its value was fixed by definition.

lectual phase locking." Most metrologists are very conscious of the possible existence of such effects; indeed ever-helpful colleagues delight in pointing them out! . . . Aside from the discovery of mistakes, the near completion of the experiment brings more frequent and stimulating discussion with interested colleagues and the preliminaries to writing up the work add a fresh perspective. All of these circumstances combine to prevent what was intended to be "the final result" from being so in practice, and consequently the accusation that one is most likely to stop worrying about corrections when the value is closest to other results is easy to make and difficult to refute.[31]

But if changes in the values of constants in the past are attributed to the experimenters' psychology, then, as other eminent metrologists have observed, "this raises a disconcerting question: How do we know that this psychological factor is not equally important today?"[32] In the case of the velocity of light, however, this question is now academic. Not only is the velocity fixed by definition, but the very units in which velocity is measured, distance and time, are defined in terms of light itself.

The second used to be defined as 1/86,400 of a mean solar day, but it is now defined in terms of the frequency of light emitted by a particular kind of excitation of caesium-133 atoms. A second is 9,192,631,770 times the period of vibration of the light. Meanwhile, since 1983 the meter has been defined in terms of the velocity of light, itself fixed by definition.

As Brian Petley has pointed out, it is conceivable that:

(i) the velocity of light might change with time, or (ii) have a directional dependence in space, or (iii) be affected by the motion of the Earth about the Sun, or motion within our galaxy or some other reference frame.[33]

Nevertheless, if such changes really happened, we would be blind to them. We are now shut up within an artificial system where such changes are not only impossible by definition, but would be undetectable in practice because of the way the units are defined. Any change in the speed of light would change the units themselves in such a way that the velocity in kilometers per second remained exactly the same.

THE RISE OF PLANCK'S CONSTANT

Planck's constant, h, is a fundamental feature of quantum physics and relates the frequency of a radiation, v, with its quantum of energy, E, according to the formula $E = hv$. It has the dimensions of action (energy $\times$ time).

We are often told that quantum theory is brilliantly successful and amazingly accurate. For example: "The laws that have been found to describe the quantum world . . . are the most accurate and precise tools we have ever found for the successful description and prediction of the workings of Nature. In some cases the agreement between the theory's predictions and what we measure are good to better than one part in a billion."[34]

I heard and read such statements so often that I used to assume that Planck's constant must be known with tremendous accuracy to many places of decimals. This seems to be the case if one looks it up in a scientific handbook—so long as one does not also look at previous editions. In fact its official value has changed over the years, showing a marked tendency to increase (Figure 16).

The biggest change occurred between 1929 and 1941, when it went up by more than 1 percent. This increase was largely due to a substantial change in the value of the charge on the electron, e. Experimental measurements of Planck's constant do not give direct answers, but also involve the charge on the electron and/or the mass of the electron. If either or both of these other constants change, then so does Planck's constant.

I have already mentioned Millikan's work on the charge on the electron in the Introduction to Part 3, and this turned out to be one of the roots of the trouble. Even though other researchers found substantially higher values, they tended to be disregarded. "Millikan's great renown and authority brought about the opinion

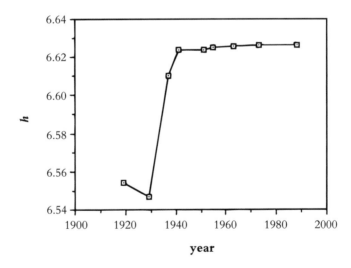

FIGURE XVI

Best values of Planck's constant (h) from 1919 to 1988.

that the question of the magnitude of e had practically got its definitive answer."[35] For some twenty years Millikan's value predominated, but evidence went on building up that e was higher. As Richard Feynman has expressed it:

> It's interesting to look at the history of measurements of the charge on the electron after Millikan. If you plot them as a function of time, you find that one is a little bigger than Millikan's, the next one's a little bit bigger than that, and the next one's a little bit bigger than that, until finally they settle down to a number that is higher. Why didn't they discover that the new number was higher right away? It's a thing that scientists are ashamed of—this history—because it's apparent that people did things like this: When they got a number that was too high above Millikan's, they would look for and find a reason why something must be wrong. When they got a number closer to Millikan's value they didn't look so hard. And so they eliminated the numbers that were too far off, and did other things like that.[36]

In the late 1930s, the discrepancies could no longer be ignored, but Millikan's high-prestige value could not simply be abandoned either; instead it was corrected by using a new value for the viscosity of air, an important variable in his oil-drop technique, bringing it into alignment with the new results.[37] In the early 1940s, even higher values of e led to a further upward revision of the official figure. Sure enough, reasons were found for correcting Millikan's value yet again, raising it to agree with the new value.[38] Every time e increased, so Planck's constant had to be raised as well.

Interestingly, Planck's constant continued to creep upwards from the 1950s to the 1970s (Table 3). Each of these increases exceeded the estimated error in the previously accepted value. The latest value shows a slight decline.

PLANCK'S CONSTANT FROM 1951 TO 1988 (REVIEW VALUES)

Author	Date	h ($\times$ 10^{-34} joule seconds)
Bearden and Watts	1951	6.623 63 $\pm$ 0.000 16
Cohen *et al.*	1955	6.625 17 $\pm$ 0.000 23
Condon	1963	6.625 60 $\pm$ 0.000 17
Cohen and Taylor	1973	6.626 176 $\pm$ 0.000 036
Cohen and Taylor	1988	6.626 075 5 $\pm$ 0.000 004 0

Several attempts have been made to look for changes in Planck's constant by studying the light from quasars and stars assumed to be very distant on the basis of the red shift in their spectra. The idea was that if Planck's constant has changed, the properties of the light emitted billions of years ago should be different from more recent light. Little difference was found, leading to the seemingly impressive conclusion that h varies by less than 5 parts in 10^{13} per year. But critics of such experiments have pointed out that these constancies are inevitable, since the calculations depend on the implicit *assumption* that h is constant; the reasoning is circular.[39] (Strictly speaking, the starting assumption is that the product hc is constant; but since c is constant by definition, this amounts to assuming the constancy of h.)

FLUCTUATIONS IN THE FINE-STRUCTURE CONSTANT

One of the problems of looking for changes in a fundamental constant is that if changes are found in the constant, then it is difficult to know whether it is the constant itself that is changing, or the units in which it is measured. However, some of the constants are dimensionless, expressed as pure numbers, and hence the question of changes in units does not arise. One example is the ratio of the mass of the proton to the mass of the electron. Another is the fine-structure constant. For this reason, some metrologists have emphasized that "secular changes in physical 'constants' should be formulated in terms of such numbers."[40]

Accordingly, in this section I look at the evidence for changes in the fine-structure constant, α, formed from the charge on the electron, the velocity of light, and Planck's constant according to the formula $\alpha = e^2/2hc\epsilon_0$, where ϵ_0 is the permittivity of free space. It gives a measure of the strength of electromagnetic interactions, and is sometimes expressed as its reciprocal, approximately 1/137. This constant is treated by some theoretical physicists as one of the key cosmic numbers that a Theory of Everything should be able to explain.

Between 1929 and 1941 the fine-structure constant increased by about 0.2 percent, from 7.283×10^{-3} to 7.2976×10^{-3}.[41] This change was largely attributable to the increased value for the charge on the electron, partly offset by the fall in the speed of light, both of which I have already discussed. As in the case of the other

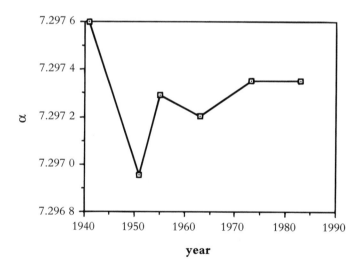

FIGURE XVII

Best values of the fine-structure constant (α) from 1941 to 1983.

constants, there was a scatter of results from different investigators, and the "best" values were combined and adjusted from time to time by reviewers. The progress of these consensus values from 1941 onwards is shown in Figure 17. As in the case of other constants, the changes were generally larger than would be expected on the basis of the estimated errors. For example, the increase from 1951 to 1963 was twelve times greater than the estimated error in 1951 (expressed as the standard deviation); the increase from 1963 to 1973 was nearly five times the estimated error in 1963. The relevant figures are shown in Table 4:

TABLE FOUR

THE FINE-STRUCTURE CONSTANT FROM 1951 TO 1973

Author	Date	α ($\times 10^{-3}$)	
Bearden and Watts	1951	7.296 953	$\pm$ 0.000 028
Condon	1963	7.297 200	$\pm$ 0.000 033
Cohen and Taylor	1973	7.297 350 6	$\pm$ 0.000 006 0

Several cosmologists have speculated that the fine-structure constant might vary with the age of the universe,[42] and attempts have been made to check this possibility by analyzing the light from stars and quasars, assuming that their distance is proportional to the redshift of their light. The results suggest that there has been little or no change in this constant.[43] But as with all other attempts to infer the constancy of constants from astronomical observations, many assumptions have to be made, including the constancy of other constants, the correctness of current cosmological theories, and the validity of red-shifts as indicators of distance. All of these assumptions have been and still are being questioned by dissident cosmologists and astrophysicists.[44]

Do Constants Really Change?

As we have seen with the four examples above, the empirical data from laboratory measurements reveal all sorts of variations as time goes on. Similar variations are found in the values of the other fundamental constants. These do not trouble true believers in constancy, because they can always be explained in terms of experimental error of one kind or another. Because of continual improvements in techniques, the greatest faith is always placed in the latest measurements, and if they differ from previous ones, the older ones are automatically discredited (except when the older ones are endowed with a high prestige, as in the case of Millikan's measurement of *e*). Also, at any given time, there is a tendency for metrologists to overestimate the accuracy of contemporary measurements, as shown by the way that later measurements often differ from earlier ones by amounts greater than the estimated error. Alternatively, if metrologists are estimating their errors correctly, then the changes in the values of the constants show that the constants really are fluctuating. The clearest example is the fall in the speed of light from 1928 to 1945. Was there a real change in the course of nature, or was it due to a collective delusion among metrologists?

So far there have been only two main theories about the fundamental constants. First, they are truly constant, and all variations in the empirical data are due to errors of one kind or another. As science progresses, these errors are reduced. With ever-increasing precision we come closer and closer to the constants' true values. This is the conventional view. Second, several theoretical physicists have speculated that one or more of the constants may vary in some smooth and regular manner with the age of the universe, or over astronomical distances. Various tests of these ideas using astronomical observations seem to have ruled out such changes. But these tests beg the question. They are founded on the assumptions that they set out to prove: that constants are constant, and that present-day cosmology is correct in all essentials.

There has been little consideration of the third possibility, which is the one I am exploring here, namely the possibility that constants may fluctuate, within limits, around average values which themselves remain fairly constant. The idea of changeless laws and constants is the last survivor from the era of classical physics in which a regular and (in principle) totally predictable mathematical order was supposed to prevail at all times and in all places. In reality, we find nothing of the kind in the course of human affairs, in the biological realm, in the weather, or even in the heavens. The chaos revolution has revealed that this perfect order was a beguiling illusion.[45] Most of the natural world is inherently chaotic.

The fluctuating values of the fundamental constants in experimental measurements seem just as compatible with small but real changes in their values, as they are with a perfect constancy obscured by experimental errors. I now propose a simple way of distinguishing between these possibilities. I concentrate on the gravitational constant, because this is the most variable. But the same principles could be applied to any of the other constants too.

An Experiment to Detect Possible Fluctuations in the Universal Gravitational Constant

The principle is simple. At present, when measurements are made in a particular laboratory, the final value is based on an average of a series of individual measurements, and any unexplained variations between these measurements are attributed to random errors. Clearly, if there were real underlying fluctuations, either owing to changes in the earth's environment or to inherently chaotic fluctuations in the constant itself, these would be ironed out by the statistical procedures, showing up simply as random errors. As long as these measurements were confined to a single laboratory, there would be no way of distinguishing between these possibilities.

What I propose is a series of measurements of the universal gravitational constant to be made at regular intervals—say monthly—at several different laboratories all around the world, using the best available methods. Then, over a period of years, these measurements would be compared. If there were underlying fluctuations in the value of G, for whatever reason, these should show up at the various locations. In other words, the "errors" might show a correlation—the values might tend to be high in some months and low in others. In this way, underlying patterns of variation could be detected that could not be dismissed as random error.

It would then be necessary to look for other possible explanations that did not involve a change in G, including possible changes in the units of measurement. How these inquiries would turn out is impossible to foresee. The important thing is to start looking for correlated fluctuations. And precisely because fluctuations are being looked for, there is more chance of finding them. By contrast, the current theoretical paradigm leads to a sustained effort by everyone concerned to iron out variations, because constants are assumed to be truly constant.

Unlike the other experiments proposed in this book, this one would involve a fairly large-scale international effort. Even so, the budget would not need to be huge if it took place in established laboratories already equipped to make such measurements. And it is even possible that it could be done by students. Several inexpensive methods for determining G have been described, based on the classical method of Cavendish using a torsion balance, and an improved student method has recently been developed which is accurate to 0.1 percent.[46]

One of the advantages of the continual improvement in precision of metrological techniques is that it should become increasingly feasible to detect small changes in the constants. For example, a far greater accuracy in measurements of G should be possible when experiments can be done in spacecraft and satellites, and appropriate techniques are already being proposed and discussed.[47] Here is an area where a big question really would need big science.

But there is in fact one way that this research could be done on a very low budget to start with: by examining the existing raw data for measurements of G at various laboratories over the last few decades. This would require the cooperation of the scientists concerned, because raw data are kept in scientists' notebooks and laboratory files, and many scientists are reluctant to allow others access to these private records. But given this cooperation, there may already be enough data to look for worldwide fluctuations in the value of G.

The implications of fluctuating fundamental constants

would be enormous. The course of nature could no longer be imagined as blandly uniform; we would recognize that there are fluctuations at the very heart of physical reality. And if different fundamental constants varied at different rates, these changes would create differing qualities of time, not unlike those envisaged by astrology, but with a more radical basis.

CHAPTER SEVEN

THE EFFECTS OF
EXPERIMENTERS'
EXPECTATIONS

SELF-FULFILLING PROPHECIES

Frequently things turn out just as expected or prophecied, not because of a mysterious knowledge of the future but because people's behavior tends to make the prophecy come true. For example, a teacher who predicts that a student will fail may treat the student in ways that make failure more likely, thus fulfilling the original prophecy. The tendency for prophecies to be self-fulfilling is well known in the realms of economics, politics, and religion. It is also a matter of practical psychology. Various ways of using these powers are the bases of countless self-help books, showing how avoiding negative attitudes and adopting positive ones help to bring about remarkable successes in politics, business, and love. Likewise confidence and optimism play an important part in the practice of medicine and healing—and in sports, fighting, and many other activities.

However one chooses to interpret it, positive and negative expectations often influence what actually happens. Self-fulfilling prophecies are commonplace. So how does this apply to science?

Many scientists carry out experiments with strong expectations about the outcome, and with deep-rooted assumptions about what is and what is not possible. Can their expectations influence their results? The answer is yes.

First, expectations affect the kinds of questions that are asked in experiments. And these questions in turn shape what kinds of answers will be found. This is explicitly acknowledged in quantum physics, where the design of the experiment determines what kind of outcome is possible; for example, whether the answer will be in wave or particle form. But this principle is perfectly general. "The structure of the examination is like a stencil. It determines how much of the total truth will appear and what pattern it will suggest."[1]

Second, experimenters' expectations affect what they observe, giving a tendency to see what they want to see and to ignore what they do not want to see. This tendency can lead to unconscious biases in observation and in the recording and analysis of data, to the dismissal of unfavorable results as errors, and to a very selective publication of results, as I discussed in the Introduction to Part 3.

Third, and more mysteriously, experimenters' expectations may affect what actually happens. Just *how* mysterious this process might be is the question this chapter explores.

EXPERIMENTER EFFECTS

A pioneering piece of industrial research, carried out at the Hawthorne plant of the Western Electric Company in Chicago in 1927–9, has become familiar to generations of students of social psychology. It revealed what is now generally known as the "Hawthorne effect."[2] This study was designed to find out the effects on productivity of various changes in rest periods and refreshments. But, to the surprise of the investigators, output increased by about 30 percent irrespective of the particular experimental treatments. The attention they were being given had a greater effect on the workers than the particular physical conditions they were working under.

The Hawthorne effect may play a part in many kinds of research, at least in psychology, medicine, and animal behavior. Investigators affect the subjects of their investigation merely by paying attention to them. Moreover, they may not only have a general influence, owing to their attention and interest, but also a specific influence on the way their subjects behave. In general, subjects tend to behave in accordance with experimenters' expectations.

The tendency for experiments to yield the expected results is known as the "experimenter effect," or more precisely the "experimenter expectancy effect." Most researchers in the behavioral and medical sciences are well aware of this tendency and try to guard against it by the use of "blind" methodologies. In "single-blind" experiments, the subjects do not know what treatment they are being subjected to. In "double-blind" experiments, the experimenters do not know either. The treatments are coded by a third party, and the experimenter does not know the code until after the data have been collected.

Important though experimenter effects are in research on human beings and animals, no one knows how widespread they are in other fields of science. The conventional assumption is that experimenter effects are widely enough recognized already, and are confined to animal behavior, psychology, and medicine. They are largely ignored in other areas of science, as can easily be seen by visiting a scientific library and looking through the journals in different fields. In research in biology, chemistry, physics, and engineering, double-blind methods are rarely, if ever, employed. Scientists in these fields are generally innocent of the possibility that experimenters might unconsciously affect the systems they are studying.

Lurking in the background is the alarming thought that much of established science may reflect the influence of the experimenters' expectations, even through psychokinetic or other paranormal influences. These expectations may not only include those of individual investigators but also the consensus among their peers. Scientific paradigms, models of reality shared by professionals, have a great influence on the general pattern of expectation and could influence the outcome of countless experiments.

It is sometimes suggested, in a joking way, that nuclear physicists do not so much discover new subatomic particles as invent them. To start with, the particles are predicted on theoretical grounds. If enough professionals believe they are likely to be found, costly accelerators and colliders are built to look for them. Then, sure enough, the expected particles are detected, as traces in bubble chambers or on photographic films. The more often they are detected, the easier they become to find again. A new consensus is established: they exist. The success of this investment of hundreds of millions of dollars then justifies yet further expense on even bigger atom smashers to find yet more predicted particles, and so on. The only limit seems to be set not by nature herself but by the willingness of the U.S. Congress to go on spending billions of dollars on this pursuit.

In the physical sciences, although there has been very little empirical research on experimenter effects, there have been many sophisticated discussions of the role of the observer in quantum theory. Such observers, discussed philosophically, sound like the detached minds of idealized objective scientists. But if the active influence of the experimenter's mind is taken seriously, then many possibilities open up—even the possibility that the observer's mind may have psychokinetic powers. Perhaps "mind over matter" phenomena take place in the microscopic realm of quantum physics. Perhaps the mind can influence the probabilities of happenings which are "probabilistic," not rigidly determined in advance. This idea is the basis of much speculation among parapsychologists,[3] and is one way of trying to explain the interaction of mental and physical processes in the brain.[4]

In the realm of animal behavior, as I discuss below, there is actual experimental evidence for the effects of experimenters' expectations on the behavior of animals. But in most areas of biology, the possibility of such effects is usually ignored. An embryologist, for example, may well recognize the need to guard against biased observation and to use appropriate statistical procedures but is unlikely to take seriously the idea that his expectations can, in some mysterious way, influence the development of embryonic tissues themselves.

In psychology and medicine, experimenter effects are generally explained in terms of influences transmitted by "subtle cues." But just how subtle these cues may be is another question. It is generally assumed that they depend only on recognized forms of sensory communication, in turn dependent only on the well-known principles of physics. The possibility that they include "paranormal" influences such as telepathy and psychokinesis is not discussed in polite scientific society. I believe that it is better to face this possibility than to ignore it, and propose an investigation of experimenter effects that takes into account the possibility of "mind over matter" effects. But first it is important to consider what has already been established.

HOW PEOPLE BEHAVE AS EXPECTED

People generally behave as expected. If we expect people to be friendly they are more likely to be so than if we expect them to be hostile, and treat them accordingly. The patients of Freudian analysts tend to have Freudian dreams, while patients of Jungian analysts have Jungian dreams. There are countless examples from all realms of human experience that illustrate this principle.

Compared with the richness of personal experience and anecdotal accounts, experiments on the effects of expectation on people's behavior seem contrived and trivial. Nevertheless, they are important in that they enable this effect to be investigated empirically and brought within the realm of scientific discourse. And hundreds of experiments have indeed shown that experimenters can affect the outcome of psychological investigations, biasing them in the direction of their expectations.[5]

Here is one example. A group of fourteen psychology graduate students was given "special training" in a "new method of learning the Rorschach procedure," in which they would be asking people what patterns they saw in ink blots. Seven of them were led to believe that experienced psychologists obtained more human than animal images from their subjects. The other seven were given the same ink blots but told that they had been found by experienced psychologists to give rise to a higher proportion of animal images. Sure enough, the second group obtained significantly more animal images than the first.

Less trivial is the empirical demonstration that the effects of such expectations are not confined to short-term laboratory experiments. In schools, for instance, the way teachers treat pupils and hence the way the children learn is strongly influenced by expectations. The textbook example is called the "Pygmalion experiment," carried out in an elementary school in San Francisco by the Harvard psychologist Robert Rosenthal and his colleagues. These prestigious scientists created expectations in the teachers that certain children in their classes were about to bloom intellectually and would show remarkable gains in the current school year. The psychologists created this belief by administering a test to all the children in the school, describing it as a new technique for predicting intellectual "blooming," calling it the "Harvard Test of Inflected Acquisition." Within each class, the teacher was then given the names of the 20 percent of the children who had scored highest. In fact, it was an ordinary non-verbal intelligence test, and the names of those most likely to "bloom" were chosen at random.

At the end of the school year, when all the children were tested again with the same intelligence test, in the first grade, the "promising" children scored an average of 15.4 IQ points more than the control children; in the second grade 9.5 points more. Not only did these "promising" children tend to score better, but there was also a tendency for teachers to rate them as more appealing, adjusted, affectionate, curious, and happy. This effect showed up much less from the third grade upwards, probably because the teachers had their own expectations about the children; the expectations created by Rosenthal and his colleagues had much less effect when they had to compete with established reputations.[6] Many subsequent studies have confirmed and amplified these conclusions.[7]

A criticism leveled against Rosenthal and his colleagues was that their own strong commitment to finding experimenter effects had biased their own results. Rosenthal replied that if this were so it would merely prove his point in another way:

> We could perform a study in which we randomly divided expectancy investigators into two groups: in the first, expectancy experiments would be conducted as usual, while in the second, special safeguards would be employed so that the expectancy of the principal investigator could not be communicated to the experimenters. Suppose that the average expectancy effect for the first group was seven, and for the second, zero. We would still view this as evidence for the phenomenon of expectancy effects![8]

Although in the medical and behavioral sciences double-blind procedures are routinely employed to guard against experimenter effects, these methods are only partially effective. Some effects of expectancy still persist, and are most clearly seen in the placebo effect in medical research.

THE PLACEBO EFFECT

Placebos are treatments with no specific therapeutic value which nevertheless help to make many people better. Medical researchers have found that placebo effects are all-pervasive in medicine. If placebo effects are not controlled in therapeutic studies, the findings are generally considered unreliable. Placebo effects have been found in many conditions, including cough, mood changes, angina pectoris, headache, seasickness, anxiety, hypertension, status asthmaticus, depression, common cold, lymphosarcoma, gastric secretion and motility, dermatitis, rheumatoid arthritis, fever, warts, insomnia, and pain symptoms from a variety of sources.[9]

Much of the success of therapy through the ages can be attributed to the placebo effect, irrespective of the kind of therapy, or of the theories supporting it. And there can be no doubt that it plays a large role in modern medicine as well. A survey of a wide range of drug trials has revealed that placebos are, on average, about a third to a half as effective as specific medication—a big effect for blank pills that cost almost nothing. But placebos are not just blank pills. They can also be forms of blank counseling or psychotherapy, or even blank surgery. For example, one surgical procedure for the treatment of angina pain involved the binding of the mammary arteries. When the effectiveness of this procedure was tested, the appropriate incision was made in control patients, but no artery was bound. "Relief from the angina pain was the same among the real and sham surgery groups. In addition, both groups showed physiological changes, including reduction in the inverted T-wave of the EKG recording."[10]

So what are placebos? The history of the word itself is revealing. It is the first word of a chant in medieval funeral rites, "placebo domino"—I shall please the Lord.[11] The word was used to refer to professional mourners who were paid to "sing placebos" at the bier of the deceased in place of the family, whose role it was originally. Over the course of several centuries the connotations of the term gradually became derisive; it was used to describe flatterers, sycophants, and social parasites. It first appeared in a medical dictionary in 1785, in a pejorative sense, defined as "a commonplace method or medicine."[12]

The professional placebo singers in the Middle Ages no doubt tended to lack any specific devotion to the deceased. Nevertheless their chanting was generally believed to be of value as part of an acknowledged ritual. Modern placebos are given in a therapeutic setting, and also depend for their power on current beliefs and expectations, both of the doctor and the patient. Any method of treatment in any culture, traditional or modern, occurs in a context in which the particular techniques are viewed by the patient as plausible and the therapists as potentially effective.

Doctors are often quick to ascribe the efficacy of traditional or "unscientific" medical systems to the placebo phenomenon, and also to impute the use of placebos to other kinds of physician. But they tend to exempt their own kind of medicine. In one survey of attitudes to placebo effects, surgeons excluded surgery, internists excluded medication, psychotherapists excluded psychotherapy, and psychoanalysts excluded psychoanalysis.[13] Moreover, in medical research, placebo effects are generally regarded as a nuisance. But perhaps the negative attitudes of physicians to placebos is just as well, since it is the other side of the coin of their faith in the special efficacy of their own techniques, which therefore tend to work better—because of the placebo effect!

The largest placebo effects occur in double-blind trials in which both patients and physicians believe a powerful new treatment is being used. If the treatment is believed by the doctors to be less effective, a smaller placebo effect is obtained. In single-blind trials, in which the doctors know which patients have been given the placebo but the patients do not, placebos are still less effective. In open conditions where the patients know they are receiving placebos, the effects are smallest of all. In other words, treatments work best if they are thought to have powerful beneficial effects by both doctors and patients. Conversely, in trials where the *active* medications are labeled as placebos, the drugs give poorer clinical results.[14]

Thus lowered expectations lead to a lowered placebo effect. This is the case with new "wonder drugs" that arouse high hopes to start with, but fail to live up to expectations. This pattern was recognized by the nineteenth-century French physician Armand Trousseau, who advised his colleagues to "treat as many patients as possible with the new drugs while they still have the power to heal."[15] There are many modern examples. For instance, at one time the drug chlorpromazine was hailed for its efficacy in treating schizophrenia, but then faith in its powers gradually waned. In successive trials it was found to be less and less effective. The effects of placebos declined in parallel. "Perhaps as the investigators began to realize that the new 'wonder drug' was not as powerful as they had hoped, their expectations, and possibly their interest in the patients declined."[16] Here is another particularly striking example, from the 1950s:

> A man with advanced cancer was no longer responding to radiation treatment. He was given a single injection of an experimental drug, Krebiozen, considered by some at the time to be a "miracle cure" (it has since been discredited). The results were shocking to the patient's physician, who stated that his tumors "melted like snowballs on a hot stove." Later the man read studies suggesting the drug was ineffective, and his cancer began to spread once more. At this point his doctor, acting on a hunch, administered a placebo intravenously. The man was told the plain water was a "new, improved" form of Krebiozen. Again, his cancer shrank away dramatically. Then he read in the newspapers the American Medical Association's official pronouncement: Krebiozen was a worthless medication. The man's faith vanished, and he was dead within days.[17]

The same principles apply to medical research itself. Believers and non-believers in new forms of treatment tend to obtain very different results: "Quantitatively, the pattern is consistent. The initial 70 to 90 percent effectiveness in the enthusiasts' reports [decreased] to 30 to 40 percent 'baseline' placebo effectiveness in the skeptics' reports."[18]

A remarkable feature of placebos is that patients not only benefit from them, but also exhibit toxic responses or side effects. In one survey of sixty-seven double-blind drug trials involving 3,549 patients, 29 percent of the patients showed various side effects while they were being treated with the placebo, including anorexia, nausea, headache, dizziness, tremor, and skin rash.[19] The side effects were sometimes so severe that they required additional medical intervention. Moreover, they showed a relationship to the doctors' or patients' expectations about the active drug being used in the trial.[20] For example, in a large-scale double-blind trial of oral contraceptives, 30 percent of the women who were administered the placebo reported decreased sex drive, 17 percent increased headache, 14 percent increased menstrual pain, and 8 percent increased nervousness and irritability.[21]

Just as the power of blessings is mirrored by the power of curses, so the beneficial effects of placebos are mirrored by the negative effects of procedures expected to bring about harm, technically known as "negative placebos" or "nocebos." Spectacular examples in Africa, Latin America, and elsewhere are known to anthropologists as "voodoo deaths," brought about by belief in the power of bewitchment. Less spectacular nocebo effects have also been demonstrated in laboratory experiments, as in a study in which subjects were told that a mild electric current was being passed through their head by means of applied electrodes, and warned that this might give rise to a headache. Although there was in fact no current, two-thirds of the subjects developed headaches.[22] Both placebos and nocebos depend on prevailing cultural beliefs, including the belief in scientific medicine. "Simply put, belief sickens; belief kills; belief heals."[23]

THE INFLUENCE OF EXPECTANCY ON ANIMALS

Animals respond to different people differently, as every pet owner and animal trainer knows. They recognize people they are used to, and tend to be on their guard with strangers. They seem to sense whether people are friendly, gauge their fear or confidence, and respond to their expectations. From a commonsense point of view, based on everyday experience, it is hardly surprising that scientists who do experiments on animals have a personal influence on the animals. The experimenters' attitudes and expectations affect the animals they work with.

The classical experiments on the effects of experimenters' expectations on animals were carried out in the 1960s by Robert Rosenthal and his colleagues. They used students as experimenters and rats as subjects. The rats came from a standard laboratory strain, but were divided at random into two groups, labeled "Maze-Bright" and "Maze-Dull." The students were told that these animals were the products of generations of selective breeding at Berkeley for good and poor performance in standard mazes. The students naturally expected the "bright" rats to learn quicker than the "dull" ones. Sure enough, this is what they found. Overall the "bright" rats made 51 percent more correct responses and learned 29 percent faster than the "dull" rats.[24]

These findings have been confirmed in other laboratories and with other kinds of learning.[25] Comparable experimenter effects have even been observed with flatworms, lowly creatures that live in mud at the bottom of ponds and in similar aquatic environments. In one such study, a sample of essentially identical *Planaria* worms was divided into two groups, one of which was described as a strain showing few head turns and body contractions ("low-response-producing worms"), and the other as a frequent turner and contracter ("high-response-producing worms"). With these expectations in mind, the experimenters found on average five times more head turns and twenty times more contractions in the "high-response-producing" worms.[26]

These expectancy effects, like those in Rosenthal's rat experiment, were shown by undergraduate students, who may be especially prone to see, or even to pretend to see, what they are told to expect. Seasoned observers might generally show smaller expectancy effects. This was the case, for example, when more experienced researchers were working with *Planaria*. The number of contractions in "high-response-producing" *Planaria* was found to be two to seven times greater than in "low-response-producing" worms, compared with the average of twenty times greater found by undergraduates. Nevertheless, a two- to sevenfold increase is still a large effect, and obviously introduces a serious bias into the results.

On the other hand, experienced observers may be strongly

committed to particular views of reality, directly or indirectly resulting in greater expectancy effects than those found among novices with less personal commitment to particular theories. They may create a climate of expectation among their colleagues and technicians, and this in turn may influence the way their animals behave.

Although expectancy effects were first systematically investigated in the 1960s and have now been demonstrated in hundreds of special studies,[27] the general principle is by no means new. For example, Bertrand Russell, writing with his customary wit and clarity, put it as follows in 1927:

> The manner in which animals learn has been much studied in recent years, with a great deal of patient observation and experiment. . . . One may broadly say that all the animals that have been carefully observed have behaved so as to confirm the philosophy in which the observer believed before his observations began. Nay, more, they have all displayed the national characteristics of the observer. Animals studied by Americans rush about frantically, and with an incredible display of hustle and pep, and at last achieve the desired result by chance. Animals observed by Germans sit still and think, and at last evolve the solution out of their inner consciousness.[28]

EXPERIMENTER EFFECTS IN PARAPSYCHOLOGY

Experimenter effects are well known to parapsychologists, for several reasons. First, it has long been known to experienced researchers that subjects tend to show more psychic powers when they are feeling relaxed, and in a positive and enthusiastic atmosphere. If they are anxious, uncomfortable, or treated in a formal and detached way by the scientific investigators, they do not perform so well. In fact they may show no significant psychic powers at all, no "psi effects," in the jargon of parapsychology.

Second, it is a matter of common observation among researchers in this field that subjects who show considerable psychic abilities often tend to lose them when strangers come into the room as observers. The pioneering parapsychologist J. B. Rhine actually quantified this effect in a series of trials with a gifted subject, Hubert Pearce, having noticed that when someone called in to see Pearce at work his scores at once dropped down. "We began to take down evidence, sometimes inviting a visitor for that purpose, sometimes availing ourselves of a casual caller. We recorded the time of entrance and exit on 7 visitors, one being present twice. They all produced a drop in Pearce's scoring."[29]

The off-putting effect of strangers is particularly strong when the strangers are sceptical, especially if they are hostile to the experiment itself or to the people involved. However, if strangers are friendly, and especially if they help in some way in the experiment, rather than behaving as detached observers, subjects get used to them and psi scores rise again.[30] Sceptics usually take the failure of parapsychological tests in the presence of sceptics to mean that psychic powers cannot be detected under rigorous scientific conditions, and therefore don't really exist. But the negative effects of sceptics may well be due to their off-putting presence and negative expectations, mediated by subtle and not-so-subtle cues.

Third, it is well known among parapsychologists that some experimenters consistently obtain positive results in their research, while others do not. This effect was systematically investigated in the 1950s by two British researchers. One, C. W. Fisk, a retired inventor, consistently obtained significant results in his experiments. The other, D. J. West, later to become Professor of Criminology at Cambridge, was usually unsuccessful in detecting psychic phenomena. In these experiments, each investigator prepared half of the test items, and scored them at the end. The subjects did not know that two experimenters were involved, nor did they meet them; they received the test items through the post and also returned them by mail. The results from Fisk's half of the experiment showed highly significant effects for clairvoyance and psychokinesis. West's data showed no deviation from chance. They concluded that West was "a jinx."[31]

Fourth, in research on psychokinesis it has repeatedly been found that experimenters who find significant effects are themselves good subjects. For example, Helmut Schmidt, the inventor of the Schmidt machine, a random number generator whose output can apparently be affected by willing certain patterns to emerge, has found that he is often his own best subject.[32] One investigator, Charles Honorton, has even shown that psychokinetic effects on random number generators by the subjects in his experiments are more due to himself than to his subjects.[33] The subjects showed psychokinetic powers when he was present; and he himself showed them when he was acting as the experimental subject. But the psi effect was lost when he was not present and the subjects were tested by another experimenter. Honorton and his colleague Barksdale concluded that such effects showed that "traditional boundaries between subjects and experimenters cannot be easily maintained." They interpreted their results as a "psi-mediated experimenter effect."[34]

The implications of such experimenter effects are staggering. If parapsychologists can bring about psi-mediated experimenter effects, whether intentionally or not, through their influence over their subjects, even at a distance (as in the Fisk-West experiments), then the conventional separation between experimenters and the subjects of their investigation breaks down. Moreover, if people can influence physical events such as radioactive decay, then the conventional separation between mind and matter breaks down too. But then why should psi-mediated experimenter effects be confined to parapsychology? Might they not occur in many other fields of science?

How Paranormal Is Normal Science?

There is a good reason for the conventional taboo against parapsychology, making it a kind of outcast from established science. The existence of psychic phenomena would seriously endanger the illusion of objectivity. It would raise the possibility that many empirical results in many fields of science reflect the expectations of the experimenters through subtle unconscious influences. Ironically, the orthodox ideal of passive observation may well provide excellent conditions for paranormal effects:

> An experimenter preparing his apparatus, getting his animals ready, and then leaving them with some feeling of assurance that the experiment will run and the animals will appropriately "do their thing" cannot but remind us of certain aspects of magic, ritual, or perhaps petitionary prayer. Something is done with the confidence that it will produce a desired result, and the participant, once he has done this, psychologically puts a distance between himself and the outcome. He is not trying to *make* things happen, but just trusts that they will. . . . Such circumstances may provide an optimum opportunity for psychokinetic intervention.[35]

This possibility has indeed been raised in a paper in *Nature* entitled "Scientists confronting the paranormal," by the physicist David Bohm and others. They noted that the relaxed conditions necessary for the appearance of psychokinetic phenomena are also those most fruitful for scientific research in general. Conversely, tension, fear, and hostility tend not only to inhibit psi effects, but also to influence experiments in the so-called hard sciences too. "If any of those who participate in a physical experiment are tense and hostile, and do not really want the experiment to work, the chances of success are greatly diminished."[36]

The defenders of orthodoxy generally reject or ignore the possibility of paranormal influences under any circumstances. The task of keeping science psi-free is undertaken by organized groups of Skeptics. These scientific vigilantes continually challenge any evidence for psi effects, rejecting it on one or more of the following grounds:

1. Incompetent experimentation.
2. Selective observation, recording, and reporting of data.

3. Unconscious or conscious deception.
4. Experimenter effects mediated by subtle cues.

Skeptics are right to point out these possible sources of error in parapsychological research. But the same sources of bias are present in orthodox research as well. The very fact that parapsychological research is subject to such critical scrutiny makes researchers in this field unusually conscious of the effects of expectation. Ironically, it is in conventional, uncontroversial fields of research that the influences of experimenters' expectations are most likely to pass undetected.

The evidence for experimenter effects in medicine and the behavioral sciences is undeniable. And that is why "subtle cues" take on such an important explanatory role. Almost everyone agrees that subtle cues such as gestures, eye movements, body posture, and odors can influence people and animals. Skeptics are very keen on emphasizing the importance of such cues, and rightly so. A favorite example showing the importance of subtle communication is the story of Clever Hans, a famous horse in Berlin at the turn of the century. This horse could apparently perform arithmetic in the presence of its owner by tapping a hoof on the ground to count out an answer. Fraud seemed unlikely, since the owner would allow other people (free of charge) to question the animal themselves. The phenomenon was scientifically investigated in 1904 by the psychologist Oskar Pfungst, who concluded that the horse was receiving clues from gestures made, probably unwittingly, by the owner and other questioners. Pfungst found that he could get the horse to give the correct answer simply by concentrating his attention on the number, though he was not aware of making any movement that would give the number away.[37]

No one denies that subtle cues from experimenters, passing through normal sensory channels, can affect people and animals. Skeptics claim that such influences may explain many examples of seemingly telepathic communication. But granted all this, the possibility remains that *both* subtle sensory cues *and* "paranormal" influences play a part.

The story of Pfungst's investigation of Clever Hans has been told again and again to generations of psychology students. What is less well known is that after Pfungst's investigation, described in his book on Clever Hans published in 1911, further studies on horses with similar mathematical powers showed that more was involved than subtle sensory cues. For example, when Maurice Maeterlinck investigated the famous calculating horses of Elberfeld, he concluded that they were somehow reading his mind, rather than responding to subtle sensory cues. After a series of increasingly stringent tests, he finally thought of one which "by virtue of its very simplicity, could not be exposed to any elaborate and far-fetched suspicions." He took three cards with numbers on them, shuffled them without looking at them, and placed them face down on a board where the horse could see only their blank backs. "There was therefore, at that moment, not a human soul on earth who knew the figures." Yet, without hesitation, the horse rapped out the number the three cards formed. This experiment succeeded with the other calculating horses too "as often as I cared to try it."[38] These results go even beyond the possibility of telepathy, since Maeterlinck himself did not know the answers when the horses were tapping them out. They imply either that the horses were capable of clairvoyance, directly knowing what was on the cards, or precognition, knowing the number that would be in Maeterlink's mind when he later turned the cards over.

For more than eighty years, the story of Clever Hans and Pfungst has been told and retold as a triumph of scepticism. It has taken on a mythic significance, enabling seemingly paranormal effects to be explained in terms of subtle cues. But what if some of the subtle cues are themselves paranormal? There is a taboo against even discussing this possibility, let alone investigating it. Nevertheless, the possible importance of parapsychological influences was suggested to Rosenthal by one of his colleagues at Harvard right at the outset of his research on experimenter effects:

Had I the wit or courage to do so, I could easily have conducted a study in which experimenters with varying expectations for their subjects' responses were prevented from having sensory contact with those subjects. My prediction, then and now, was (and would be) that under these conditions no expectancy effects could occur. But I never did the study.[39]

Maybe if someone actually did this study, Rosenthal's prediction would turn out to be wrong. Maybe some of the effects of experimenters' expectations are indeed paranormal. Such subtle influences would not be opposed to subtle cues; they would usually work along with them, and operate just as unconsciously.

Although experimenter effects are well recognized in the medical and behavioral sciences, the fact that they are explained—or explained away—in terms of "subtle cues" prevents them from being taken very seriously in other fields of investigation such as biochemistry. Whereas a person or a rat might pick up a scientist's expectations and respond accordingly, an enzyme in a test tube would not be expected to respond to subtle body language, unconscious facial gestures, etc. Of course, there is a general recognition of the possibility of biased observation, but this is not a result of any actual influence on the experimental system itself. The scientist may "see" a difference that fits his or her expectancy, but the difference is supposed to be only in the eye of the observer, not in the material studied.

Nevertheless, all this is merely an assumption. There has been practically no research on the influence of experimenters' expectations in fields of science such as agriculture, genetics, molecular biology, chemistry, and physics. Since the material studied is assumed to be immune from such influences, precautions against them are assumed to be unnecessary. Except in the behavioral sciences and in clinical research, double-blind procedures are rarely employed.

I now suggest a variety of tests to explore the possibility that experimenter effects may be far more widespread than previously thought.

Experiments on Possible Paranormal Experimenter Effects

In looking for experimenter effects, I think it is best to start with situations where the phenomena show an inherent variability, an inherent indeterminism, allowing scope for the biasing effects of expectancy. This is certainly the case with human and animal behavior, where expectancy effects have been so clearly demonstrated. I would not expect physical systems with a high degree of uniformity and predictability to show much scope for biasing effects, for example the dynamics of billiard balls (although even here, in a hotly contested game of billiards, a player might be highly motivated to affect the outcome of impacts and collisions, and could conceivably bring into play unconscious psychokinetic powers).

In fact variable, statistical results are the norm in most fields of social and biological research, including sociology, ecology, veterinary medicine, agriculture, genetics, developmental biology, microbiology, neurophysiology, immunology, and so on. And so they are in quantum physics, where probabilities are of the essence. There are many areas of the physical sciences too where inherent variability is very apparent, as in crystallization processes—for example, every snowflake is different. And even the most mechanistic of systems, mass-produced machines, are variable. Their tendency to break down, for example, is measured statistically, as in the "reliability" figures for different brands published in consumer surveys. And almost everyone has heard of "lemons," individual cars or other machines which are exceptionally unreliable—in extreme cases even said to be "jinxed."

What I am proposing is a general type of experiment that can be conducted in many fields of enquiry. The experimental design follows Rosenthal's standard procedure but is extended to other areas which are so far unexplored. The purpose is to find out which systems are susceptible to the influence of experimenters' expectations, and to compare the susceptibilities of different systems. Here are two extreme examples.

First, students are given two samples of radioactive tracers, of the kind routinely used in biochemical and biophysical research, and led to believe that one is more radioactive than the other. In fact, both are the same. They then determine the levels of radioactivity following standard laboratory procedures, with automatic Geiger or scintillation counters. Do they tend to find more radioactivity in the samples where they expect it?

In the second example, in the field of consumer research, volunteers are given samples of a standard product, say an automatic camera, and told they are taking part in an experiment to study the "Monday morning" effect, whereby an unusually high proportion of "lemon" cameras are produced on Monday mornings. Half the cameras, drawn at random from a normal consignment, are labeled "Monday Morning Sample." The others are labeled "Reliable Control." The experiment is designed so that both lots of cameras are used to an equal extent under comparable conditions, and the volunteers are asked to report regularly on any problems they have encountered. Do the "Monday morning" cameras tend to show more defects?

I would guess that the camera experiment might show a bigger expectancy effect than the radioactivity experiment. There are more ways that people's expectations could affect the results—for instance, they may be more on the look-out for faults with the "Monday morning" cameras, or treat them with less respect, handling them more roughly. There would also be the possibility of unconscious paranormal influences; for example their negative expectations about the "Monday morning" cameras might somehow put a "jinx" on them. But even the radioactivity experiment leaves scope for several kinds of influence, including conscious or unconscious errors in preparing the sample for radioactive analysis, and a psychokinetic influence on the process of radioactive decay itself, or on the operation of the measuring instrument. If these experiments did in fact provide positive evidence for expectancy effects, further research could then be designed to tease apart the possibilities, separating possible paranormal effects from other sources of bias.

Here are some more examples of experiments of this general type.

1. A crystallization experiment

Many compounds do not crystallize readily even from supersaturated solutions; there may be delays of hours, days, or even weeks before crystals appear. However, crystallization can be initiated by putting in "seeds" or "nuclei" around which the crystals can form. In this experiment, students are given a supersaturated solution of a hard-to-crystallize substance and also two samples of a fine powder, one described as a "nucleation enhancer," made by a special seed-enrichment process, and the other as an "inert control." In fact the two powders are identical. To each of several identical containers containing a fixed amount of supersaturated solution, the students add a small, defined amount of "nucleation enhancer"; to an equal number of identical containers with the same amount of supersaturated solution, they add the same amount of "inert control" powder. They examine the samples at regular intervals, recording which ones have crystallized. Do the samples they expect to crystallize sooner show a tendency to do so?

2. A biochemical experiment

Students in a biochemistry practical class are given two samples of a particular enzyme. One is described as having been treated with an inhibitor which partially blocks its activity; the other is described as the untreated control. In fact both samples are identical. They measure the enzyme activity, using standard biochemical techniques. Does the "inhibited" enzyme tend to show lower activity than the "control"?

3. A genetic experiment

Students in a genetics practical class are given seeds of a fast-growing plant species, for example *Arabidopsis thaliana,* a small plant in the mustard family commonly used for genetic research. They divide this lot of seeds into two samples. One is the control. The other is placed in a lead-shielded radiation chamber, covered with signs saying "Radioactivity—Danger," and left there for a defined period before being taken out with great care. These seeds have now supposedly been subjected to powerful mutation-inducing radiations (but in fact there is no radioactive source in the chamber). Both samples are now raised under identical conditions, and the students in due course record the number of abnormal growth forms in both samples. Do they tend to find more "mutant" forms in the "irradiated" samples?

4. Another genetic experiment

Students in another genetics practical class are given fruit flies containing mutant genes, for example mutations in *bithorax* genes giving flies containing them a tendency to produce four wings instead of two (Figure 18). Such mutations are recessive; in other words, only flies with a double dose of the mutant genes develop abnormally. First-generation hybrids between such mutant flies and normal flies appear normal. But when these hybrids are crossed with each other, they give rise to progeny which show Mendelian segregation:

most of these second-generation hybrids are normal-looking, but a minority show the mutant form to varying degrees.[40]

The students are given two samples of the normal-looking hybrid flies, drawn from the same population, but one sample is said to have an "enhancer" gene that causes the bithorax character to show a higher penetrance and expressivity in the segregating population. (In the jargon of genetics, "penetrance" means the proportion of the organisms showing the effects of the gene in question, and "expressivity" the intensity with which the effects of the gene are realized.) The other sample of hybrid flies is said to be bred from a strain with an "inhibitor" gene with the opposite effect.

The students then breed from the flies with the "enhancer" gene and the "inhibitor" gene, and carefully examine the resulting populations of flies. Do the populations with the "enhancer" gene tend to show a higher proportion

FIGURE XVIII
(A) normal fruitfly
of the species
Drosophila
melanogaster.

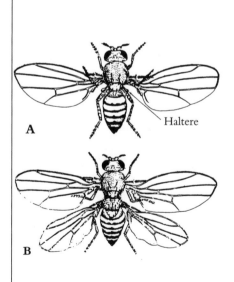

(B) A mutant fly
in which the third
thoracic segment has
been transformed in
such a way that it
duplicates the second
thoracic segment.
The balancing
organs, the halteres,
have been
transformed into
a second pair of
wings. Such flies are
known as bithorax
mutants.

of abnormal flies, and does this character tend to be more strongly expressed? (The flies in both populations should be preserved, for example in alcohol, so that they can later be re-examined independently.)

5. An agricultural experiment

Agriculture students are told that as a practical exercise they are going to carry out a trial of a promising new growth stimulator, which when sprayed on to plants at regular intervals leads to enhanced yields. They carry out a field experiment on a crop of, say, beans, using a standard design, with replicated plots and a randomized allocation of test and control treatments. Throughout the flowering and fruiting season they spray the plants in the test plots at weekly intervals with the "growth stimulator" solution, and the control plants with an equal volume of water. In fact the "growth stimulator" solution is nothing but water. On each occasion they observe the plants carefully and note any differences they can see between the plants in the test and control plots. When the crop is mature, they harvest the plants in the various plots and measure their total weight and seed yield. Do the "stimulated" plants grow better and give a higher yield than the "controls"?

There is no need to multiply examples further. Clearly the same general principles could be extended to many fields of research. Such experiments would be particularly easy to carry out, at little cost, in the context of student practical classes, with the cooperation of the regular course instructors.

DECEPTION

The only disadvantage of these experiments is that they involve deception. In this respect they follow the precedents established by Rosenthal and his colleagues, and by the use of placebos in medical research. Some people may object on ethical grounds, and I am not entirely happy with the use of deception as a means of affecting people's expectations. But I believe this kind of research can be justified because of its importance in helping to reveal the possible extent of expectancy effects in the practice of science, together with the dangers of self-deception.

However, I also believe that if such deception became commonplace it would be self-limiting. If experiments of this type give interesting and significant results, if further research on the topic became widespread, and if the results were well publicized, students would probably become increasingly aware of the possibility that they were sometimes being deceived by their instructors. They might then tend to be more sceptical about what they are told to expect, and hence less prone to expectancy effects. If the deliberate practice of deception makes students more aware of the effect of expectation, more on guard against it, this would be a valuable contribution to their scientific education.

The effects of the kind of deception used in these experiments may be relatively weak, because students' expectations may be lightly held, not involving much personal commitment; they are merely carrying out practical exercises that no one takes too seriously. Professional researchers, steeped in the currently accepted paradigms and with careers and reputations at stake, may show much stronger expectancy effects, and also be more prone to self-deception.

It would be fascinating to look for expectancy effects in disputed areas of science, especially in those situations in which proponents of a particular theory obtain experimental results supporting it, while their opponents get opposite results. One way of doing this would be to invite both sides in the dispute to carry out the same experiments in a neutral laboratory, under standardized conditions agreed in advance by all parties. If they still tended to obtain opposite results in accordance with their expectations, then this expectancy effect, including possible paranormal influences, could be investigated in detail in a real-life research situation.

Indeed, this idea could become the basis for a new kind of research center, combining investigations of the experimental method with a kind of scientific mediation service (perhaps even offering counseling facilities for the visiting disputants).

CONSEQUENCES

If any significant expectancy effects show up, the investigations will need to be taken further in order to find out what kinds of factors, normal or paranormal, may have been playing a part. For example, in experiment 4, if a bias appeared in the ratios of abnormal to normal flies in the populations of second-generation hybrids in accordance with the experimenters' expectations, the first thing to check would be a possible bias in the recording of data. This could be done by a third party, who would count the preserved flies "blind," not knowing which sample was which. Perhaps this check would show that the entire experimenter effect could be explained in terms of biased counting. On the other hand, perhaps it would show that only a part of the bias was introduced at this stage; it might confirm that the proportions of normal and abnormal flies really had been altered. Then there would have to be a check on the possibility that the experimenters did not preserve and count all the second-generation flies, but only a selected sample which might have been biased. But if it turned out that all the flies had been preserved, then the alteration in the ratios would begin to look like a paranormal effect.

A new experiment would be needed to resolve this question. The second experiment would be a repetition of the first, except that the experimenters would see the hybrid flies being crossed, but not be allowed to handle the flies or the fly-bottles until the second generation of flies had matured and was ready for counting. The flies would be looked after by people who did not know what expectations were being tested. If the expectancy effects still showed up after experimenters had no normal means of influencing the breeding and development of the flies, then they could be inferred to result from some paranormal influence.

The possible discovery of subtle paranormal expectancy effects in these and other fields of science would be shocking, even sensational. There would be many implications. One of the most important would be for the notion of consensus reality on which empirical science depends. Scientific data are regarded as objective if they can be replicated by independent observers. But in new and disputed areas of research there is not yet a consensus. As a consensus is established, the results of the relevant experiments come to agree increasingly well with expectations. But which is the cause, and which the effect? Are repeatable results the basis for consensus expectations, or are consensus expectations the basis for repeatable results? Perhaps both processes work together. But at least in the case of scientific education, consensus reality takes precedence.

Students spend many hours in laboratories doing practical classes, in which they perform standard experiments illustrating the fundamental principles of the prevailing paradigm. These experiments have "correct" results, namely those that conform to a well-established pattern of expectation. Nevertheless, these are not always the results that students obtain. I have had years of experience teaching in undergraduate laboratory classes, and have often been amazed by the variation in the students' results. Of course, deviant data are immediately put down to mistakes and inexperience. And students who persistently fail to get their experiments to work correctly are not regarded as promising researchers. They fare badly in their practical examinations, and are unlikely to pursue a scientific career. By contrast, professional scientists have succeeded in a long process of training and selection, in the course of which they have proved their ability to get the expected results from standard experiments. Is this success simply a matter of practical competence? Or does it also involve a subtle and unconscious ability to bring about experimenter effects in accordance with orthodox expectations?

CONCLUSIONS TO

PART THREE

If constants turn out to be variable, our understanding of nature would be radically changed. But it is unlikely that the edifice of established science would come crashing down like a house of cards. Scientists are generally quite pragmatic, and most would probably adapt to the new conditions quite easily. The current values of the "constants" would be reported regularly in journals such as *Nature,* rather like weather reports, or like stock market prices and currency fluctuations on the financial pages of newspapers. Those who needed these accurate, up-to-date values for their calculations would use them, but for most practical purposes the variations might not make much difference.

But although scientists could, no doubt, adapt to fluctuating constants in practice, the spirit of science would be much changed. The old faith in the mathematical constancy of nature would seem naive. Nature would appear to have a life of her own, to be unpredictable in detail and full of surprises—just like real life. Mathematicians might try to model the fluctuations in the constants, but their predictions might be no more accurate than mathematical models of the weather, the economy, or stock market cycles.

Likewise, if experimenter effects turned out to be far more widespread than currently assumed, most scientists would probably respond pragmatically. The use of double-blind procedures would be extended to as much of science as possible. In practice, double-blind experiments would be a nuisance to many biologists, chemists, and physicists, and would take a lot of the fun out of research. But experimental psychologists and clinical researchers have lived with this situation for decades, so their example shows that it is possible to adapt.

Even so, double-blind procedures do not entirely eliminate the influence of expectations, as the placebo effects discussed on pp. 215–219 show. The double-blind conditions mean that the experimenter is looking for the expected effect everywhere, not knowing exactly in which samples or subjects it will appear. This expectation tends to make the effect show up in the controls: patients given placebos often show the expected effects of the treatment under test, including side effects.

If experimenter effects had to be taken seriously in most branches of science, not just in medicine and psychology, a debate about their effects and an interest in their nature would probably lead to an expansion of research on the subject, and the area would quickly attract increased funding. But never again would a faith in the inherent objectivity of experimenters be possible in the naive form still prevalent today.

What about the experiments proposed in this book? Would they not also be subject to experimenter effects? Perhaps, but I think only to a limited extent. Wherever possible, the experiments involve blind procedures. Consider, for example, the experiments with pets who seem to know when their owner is about to return. The person observing them at home should observe them "blind," not knowing when the absent one is returning. If the pet, deprived of sensory cues and routines, can still anticipate the absent person's return, then there would be three possible explanations. Either the pet can do it because of a direct connection to its owner, or because it is responding to the expectations of the person observing it, and this person is picking up telepathically when the absent person is coming home; or a combination of the two.

Further research could then be done to tease apart these possibilities. The explanation in terms of a telepathy mediated experimenter effect could be investigated directly. The person at home could be tested to see how well he or she could anticipate the return of the absent one without the *pet* being present. Also, the behavior of the pet could be monitored by means of an automatically operated video camera without the *person* being present. If the pet still anticipated the returning person in the absence of a human observer, this could not be dismissed as an experimenter effect.

In the homing experiment in chapter 2, if it turned out that pigeons could find lofts that had moved large distances, the idea that they did this because of experimenters' expectations would make the effect even more mysterious, and still leave their direction-finding powers unexplained. In the experiment on termites in chapter 3, if the separated members of the colony behaved in a coordinated way, to explain this as an experimenter effect would seem highly implausible.

In chapter 6, the measurements of a constant at any given location could not feasibly be done blind, but by comparing the measurements of the constant at different locations to see if the fluctuations were correlated the effects of expectation could be minimized—as long as the researchers involved did not compare notes as the experiment was proceeding.

These examples show that practical research is still perfectly feasible even if the effects of experimenters' expectations are widespread. But the current assumption that there is a sharp separation between subject and object, between experimenter and the experimental system, would have to be abandoned.

On the other hand, it may turn out that in most fields of science experimenter expectancy effects are rare or non-existent, and that there is not even the slightest hint of any paranormal influence. This is what most scientists assume, and it is an article of faith for Skeptics. Thus this belief would for the first time have been tested empirically. An attempt would have been made to refute it, and the failure of this attempt would provide some evidence in its support.

Hence Skeptics with the courage of their convictions should welcome this research program as much as those, like myself, who think experimenter effects in conventional scientific research are possible, if not probable.

GENERAL

CONCLUSIONS

The research program suggested in this book subjects some of the most cherished assumptions of established science to the test. Seven typical "scientific" beliefs are examined. They are so widely taken for granted, so rarely questioned, that they are not even regarded as hypotheses, but more like scientific common sense. The opposite positions are simply regarded as "unscientific."

1. Pet animals can't really have uncanny powers.
2. Homing and migration are explicable in terms of known senses and physical forces.
3. Insect colonies are not superorganisms with a mysterious soul or field. No such things exist.
4. People can't really tell when they are being looked at from behind, except perhaps by means of subtle cues.
5. Phantom limbs are not really "out there" where they seem to be; they are in the brain.
6. The fundamental constants of nature are constant.
7. Competent professional scientists do not allow their beliefs to influence their data.

From the conventional scientific perspective, there is no point in wasting valuable scientific resources in examining the possibility that these assumptions might be wrong. It is not even worth wasting time thinking about them, especially when there are so many genuine scientific problems to be getting on with. These tenets are not refutable hypotheses; they are part of established science. The alternatives are simply unscientific, and there is no point in taking them seriously. You might as well take seriously the idea that the moon is made of green cheese.

If I were a betting man, I would try to involve a bookmaker in taking bets on the outcome of these seven experiments. Presumably most supporters of the established scientific worldview would put their money on the failure of these experiments to reveal any effects inexplicable by established science. But some people would wager the opposite, and the odds would give a measure of the relative strengths of punters' expectations. For example, how much would *you* bet that pets cannot really sense the return of their owners, if all conventional means of knowing are eliminated? Or how much would you bet that they can?

I cannot foresee the outcome of the experiments proposed here, but I think there is a good chance that at least some of them will yield very interesting results. I would not have written this book otherwise.

It is precisely because the kinds of research proposed in this book are taboo within institutional science that they have been so neglected. And that is why they present such amazing opportunities for discovery. We could be on the threshold of a new era of science, with an invigorating sense of freshness and discovery, of openness, and of general participation. Probably after a decade or two a new orthodoxy would become well established, an exclusive professionalism grow up again, and bureaucracy regain control. But the immediate prospect is exciting.

How could these experiments change the world? First of all, I think they could do so by helping to open up science, both in practice and in theory. The cultural effects of such a change in science would be profound. There would be a new valuation of folklore and popular belief, such as beliefs in uncanny powers of animals and the sense of being stared at. There would be a greater sense of our connection and affinity with each other and with the natural world around us. There would be a further swing against the idea that we have a right to conquer and exploit nature as we see fit, with no concern for anything but human interests, the rest of nature being treated as inanimate and mechanical. There would be great changes in education. There would probably also be a remarkable increase in the public interest in science.

Second, the experiments in Part 1 could lead to a new understanding of animal powers—and of human powers. They could provide evidence in favor of invisible connections between animals and people, between animals and their homes, and within social groups. The nature of these connections would need much further research, but it would probably go far beyond anything yet dreamed of within the established sciences. A wide variety of biological and human phenomena would need to be reinterpreted, including animal migration, the sense of direction, social bonding, and the organization of societies.

Third, the experiments in chapters 4 and 5 could lead to a new understanding of our relationship to our own bodies and the world around us, breaking down the conventional separation between mind and body, and between subject and object. The psychological, medical, cultural, and philosophical implications of this change would be profound.

Fourth, the experiments in Part 3 could undermine the conventional scientific beliefs in the constancy of nature and the objectivity of scientific research. They would bring home a point made by the philosopher of science Karl Popper in his book *The Logic of Scientific Discovery:*

Science does not rest upon solid bedrock. The bold structure of its theories rises, as it were, above the swamp. It is like a building erected upon piles. The piles are driven down from above the swamp, but not down to any natural or "given" base."[1]

It may well turn out that the constancy of the "constants of nature," long assumed to be a natural or given base for the edifice of science, is just such a pile in a swamp. So may be the assumption that experimenters' expectations do not introduce a major source of bias into research. As these foundations become increasingly unsteady, there will be a wider appreciation of the need to drive the piles deeper, or to find other kinds of support, such as floats.

Finally, whatever the outcome of these experiments, at the very least I hope that this book will serve to show that there is a great deal we do not understand. Many fundamental questions are still open. And we need to keep our minds open too.

UPDATES ON THE

SEVEN EXPERIMENTS

CHAPTER 1

PETS WHO KNOW WHEN THEIR OWNERS
ARE RETURNING

Experiments on return-anticipating dogs have already proved highly successful, as I describe in my book *Dogs That Know When Their Owners Are Coming Home*. In more than 150 videotaped tests with several different dogs, my associate Pam Smart and I have found that the animals showed their anticipatory behavior long before their owners returned home, when the person was still more than five miles away. They did so even when their owners set off at randomly selected times, and when no one at home knew when they would be returning. The dogs also anticipated their owners' arrivals when the humans were travelling in unfamiliar vehicles such as taxis. These effects were clear-cut and highly significant statistically.[1]

The results of these experiments, which provide strong evidence that the dogs were picking up their owners' intentions telepathically, have been replicated independently by skeptics who tested one of these animals, Jaytee, Pam Smart's own dog. Their data showed the same pattern of anticipatory behavior that Pam and I had repeatedly observed with this Jaytee.[2]

Anticipatory behavior in dogs is common. In random surveys of 1,200 households in Northern and Southern California,[3] Northwest England[4] and London,[5] most dog owners said their dogs anticipated the return of a member of the household. Many cat owners had also noticed such anticipatory behavior, but cats exhibited it less than dogs. On average 55 percent of dog owners said their animals anticipated the return of a member of the household, compared with 30 percent of cat owners.[6] This does not necessarily mean that cats are less sensitive than dogs; they may simply be less interested in their owners' comings and goings. A number of other species anticipate their owners' returns, including parrots.

So far the only systematic, videotaped experiments on animals' anticipation of their people's return have been carried out with dogs. There is much scope for further research on this subject, particularly with cats and parrots. For suggestions about other ways in which the unexplained powers of animals can be investigated simply and inexpensively, see appendix A in my book *Dogs that Know When Their Owners Are Coming Home*.

CHAPTER 2

HOW DO PIGEONS HOME?

In 1994, soon after this book was first published, I discussed the problem of pigeon homing with the biologist Stephen Jay Gould, the philosopher Daniel Dennett, and the neurologist Oliver Sacks on a Dutch television program called *A Glorious Accident*.[7] This discussion led to a continuing debate in Holland about the basis of pigeon navigation. As a result, through the initiative of Louis van Gasteren, a well-known filmmaker, a mobile loft experiment was set up at Utrecht University under the supervision of Dr. Wim Nuboer. The training process and tests followed the general procedures I described in chapter 2, and the results were similar.

When the loft was moved over short distances, up to 900 yards, the pigeons usually re-entered it within a few hours. Van Gasteren filmed these experiments.[8] When the loft was moved 1,200 yards, the pigeons took five days to re-enter it. This is not because they could not find it, but probably because they were too timid to go in, as in the experiments in England. When the loft was moved even farther, 2.75 miles, they did not enter it at all. Again, as in England, they were probably too frightened to enter the loft when it was in such an unfamiliar environment.[9]

The tests in England and Utrecht made it clear that the mobile loft experiment is unlikely to work on land. When the lofts were first moved, the pigeons took hours to enter them. They adapted to repeated moves over fairly short distances of up to about half a mile. But with distances of several miles, even when the pigeons found the loft, they would not go in. Thus tests with mobile lofts could not be expected to succeed if the lofts were moved many miles into an entirely unfamiliar environment.

The pigeons' fear is not too difficult to understand. Imagine you came home one day and found your house had disappeared, leaving a gap where it had been before. Astonished, you might look around and see it a hundred yards away. But you would not walk over and go straight in. You would probably look repeatedly at the place where it had been and wander around the area and search for clues that might explain its mysterious change of location. Only after watching warily for many minutes, or even hours, might you venture into the house in its new position. This is how the pigeons behaved when their loft was moved for the first time. If, however, your house kept moving to nearby places at unpredictable intervals, you might get used to it and go in fairly quickly. But imagine it had moved many miles away, to an entirely unfamiliar place. Even if you found it by climbing hills and looking for it with binoculars or discovered it by searching at random, in a totally strange environment with unfamiliar people and animals around it, you might be reluctant to enter it at all.

The only way forward was to perform the tests at sea. Fortunately, van Gasteren was able to persuade the Royal Dutch Navy to allow a pigeon experiment to be conducted on one of their main research ships, the *Tydeman*. He also persuaded a leading Dutch manufacturer to donate a pigeon loft. A retired seaman, Hans van der Vliet, who was an experienced pigeon fancier, volunteered to sail on the *Tydeman* and care for the birds. Van Gasteren was interested in pursuing this line of research because he was at the time making a documentary film about pigeons and he also had a strong personal curiosity.

Dutch pigeon fanciers donated most of the birds needed to stock the loft and the Swiss Army Pigeon Corps also gave four pairs. The ancestors of these Swiss birds had been trained over several generations to home to mobile lofts. We are grateful to Dr. Hans-Peter Lipp, of Zurich, the officer commanding pigeons in the Swiss Army, for this donation. (Sadly, the Swiss Army Pigeon Corps, the last military pigeon organization in the western world, has subsequently been disbanded.) The only item that remained to be accounted for was the pigeon feed. The Dutch navy had no budget for this item, so I paid for it myself. Fortunately the cost was, as it were, chicken feed.

The *Tydeman* left the Dutch naval port of Den Helder on March 4, 1996, and returned on October 11 of the same year. It set sail first to the Caribbean, where it called at Curaçao, then made its way across the Atlantic to the Canary Islands, off the Northwest coast of Africa, then to the Island of Madeira, and finally to Spain and back to Holland. The primary goal of the voyage was scientific and technological research.

In total, 73 young pigeons were bred on board the *Tydeman*, of which 12 stemmed from the Swiss army parents. These pigeons were trained entirely at sea while the ship was out of sight of land. This in itself was an innovation.

The biologist Geert van Oortmerssen of the University of Groningen was on board the *Tydeman* for part of its voyage and wrote a detailed report on the project.[10] He described the birds' flight over the water as follows: "It was truly fascinating to see how, after being released, they seemed to 'surf' the waves, while not showing any sign of fear of water while doing so. Sometimes they even seemed to want to land on top of the white foam formations that tend to emerge in the water turbulence behind a ship. In all cases the pigeons immediately resumed their flight after landing."[11]

While the pigeons were being trained on the Atlantic Ocean, the ship was usually either at rest or moving slowly with a maximum speed of 3 knots per hour. Sometimes the pigeons were away from the ship for several hours and in some cases for up to 10 hours, during which time the ship had moved more than 20 miles from the place where the pigeon was released. Probably the pigeons could see and recognize the ship, which was painted white, even though at this distance no human would be able to see it with the naked eye.

These observations are important in that they show that the birds were able to find and enter the loft after large displacements when they were at sea. In fact, because the ship moved more than 6,000 miles over the course of several months, their home was continually moving and they continued to enter it after training flights in very different geographical positions. This observation confirms that their reluctance to enter the loft after displacements of a few miles on land was not so much because the loft was in a new position as because it was in unfamiliar surroundings.

When the birds were released for flights from the ship in a new position, according to van Oortmerssen they usually flew off in a direction that "tended to coincide more or less with the compass direction of the spot where they had flown before. This indicates that that the pigeons must recognize the spot and tend to return to it."

On some days, some of the pigeons flew up very high before disappearing from sight at high speed. Some failed to reappear, especially when the ship was near land. It therefore seems possible that these birds were seeking land, and when they saw it, some headed for it.

All this training was designed to lead up to the key experiments, when pigeons would be taken off the *Tydeman* onto another ship. The *Tydeman* would then sail away in a randomly chosen direction for a distance of at least 40 miles—beyond the horizon—and the pigeons would then be released. Would they find the *Tydeman* and their loft? Unfortunately, this crucial test was not to happen. The period of time set aside for this experiment had to be curtailed from one week to two days because extra time was needed for testing some military sonar equipment, which was one of the voyage's main purposes.

The only available days for the pigeon experiments were September 14 and September 20, 1996. On September 14 the *Tydeman* was about 100 miles south of the island of Madeira. The water was calm and the weather fine. Some of the trained pigeons were taken away on a smaller ship, the *Middelburg*, and were liberated at various distances from the *Tydeman*. The first three birds were released at 7:08 A.M. GMT, about 1.5 miles northeast of the *Tydeman*. The first bird arrived back 20 minutes later and the second after 67 minutes; the third never returned. A second batch of pigeons was released at 7:33 A.M., about 5 miles west of the *Tydeman*. One of these birds arrived back after about 50 minutes; the others flew around in the vicinity of the *Middelburg*. At 8:15 A.M. a final batch of ten pigeons was released about 12 miles from the *Tydeman*. One arrived back at the *Tydeman* 2 hours later, but most stayed near the *Middelburg*. During the course of these tests, the *Tydeman* moved at least 20 miles south, so the birds that succeeded in finding it were not simply returning to the previous geographical position of the ship, although they might have headed in that direction to start with.

On September 20 the *Tydeman* was anchored about 16 miles from the coast of Madeira. The weather was fine, with a light wind from the west northwest. Some of the trained pigeons were transferred to another Dutch naval ship, the *Mercuur*, and released at various distances from the *Tydeman*. The first release of eighteen pigeons occurred from 2 miles away. All the pigeons were back on the *Tydeman* within 30 minutes. After being fed they were taken back to the *Mercuur*, which headed out against the wind, WNW, to anchor 5 miles from the *Tydeman*. Four pigeons were released from this position and all returned to the *Tydeman* within 15 minutes. A French frigate happened to be passing nearby, but although they flew towards it at first, they were not tempted to land on it.

The *Mercuur* then moved farther away, and at 12:10 P.M. two more birds were released at a distance of 10 miles, when the *Tydeman* was too far away to be visible to any man on board. The birds arrived back at their loft 30 minutes later. At 1:10 P.M. two more birds were released at a distance of 20 miles from the *Tydeman*, which subsequently started heading at full speed in a northeasterly direction. One of these birds reached the *Tydeman* at 6:30 P.M., when the ship had already moved at least 13 miles from its original position. Because this bird had been released upwind of the *Tydeman*, it could not have found it by smelling it. The other birds did not return.

Some of the most intriguing findings concerned birds that were not part of the formal experiments, but which disappeared for long periods after being released on training flights. One pigeon released on a training flight near Madeira on September 16 rejoined the *Tydeman* four days later, when it was heading at full speed (15 knots) towards Spain. When the pigeon returned to the ship, the *Tydeman* had moved more than 60 miles from the place where the bird was released. The bird could not have been flying continuously for four days, so presumably it had spent some of the time on land on the island of Madeira. More remarkable still is that case of a bird that was released, along with several others, in the mid-Atlantic on August 17, while the *Tydeman* was heading northeast at full speed. These birds had been fed shortly before release and set off rapidly towards the southeast. The nearest land was more than 1,000 miles away, beyond the range of a pigeon. All but one of these birds were lost; the single bird rejoined the ship 300 miles from the place at which it had been released.

Although the formal experiments we planned had to be curtailed for reasons beyond our control, this pioneering research with pigeons bred and trained at sea showed very remarkable feats of navigation on the part of some of the birds. Since we do not know what they did between their release and their return to the *Tydeman*, we cannot know how they found the ship again. An invisible connection between the pigeon and its home remains an open possibility.

In conclusion, this series of experiments with mobile lofts has made it clear that experiments on land are unlikely to succeed. Experiments at sea are much more likely to provide answers and have already given intriguing results. Marine experiments, however, are complicated to arrange and are not feasible on a do-it-yourself basis, except for people who happen to own ships.

In any future work on pigeon navigation at sea, it would be very desirable to include a radio tracking system, so that the pigeons' movements could be recorded continuously. This may not yet be technically possible, because devices small and light enough to be attached to pigeons are still in the process of development. In any case, they would be expensive. This project would require serious funding. Even so, the expense would be small compared with the budgets routinely allocated to many other fields of research.

CHAPTER 3

THE ORGANIZATION OF TERMITES

I had an opportunity to investigate termite colonies in August 1998, when I spent three weeks in Brazil, at the Buzios Ecological Institute in the Atlantic rain forest. I tried Marais's basic experiment by inserting aluminum plates into the mounds of two species of termites. Unfortunately, unlike the *Eutermes* species that Marais studied in South Africa, in one of the species I tested, the termites did not repair the damage. Those of the other species carried out some repairs, but in the short time available for my observations there was not enough rebuilding activity to come to any conclusions about how well it was coordinated on both sides of the metal plate.

I have not heard of any other investigations with termites. Several students have carried out projects with ants with promising results, but the projects were too short to reach any definite conclusions.

This is still a wide open field for research. For people not living in the tropics, the most feasible experiments to conduct are those involving ant colonies kept in separable containers.

CHAPTER 4

THE SENSE OF BEING STARED AT

Of all the experiments proposed in this book, tests on the sense of being stared at have proved to be the most popular. Thousands of people have now taken part, and by the beginning of 2002 about 50,000 trials had been completed. The results were repeatable, positive, and extremely significant statistically.

Since the first edition of this book was published I have slightly modified the test procedure. In the basic experiment, people work in pairs, one serving as subject, the other as looker. The subject sits with his or her back to the looker and wears a blindfold of the type airlines provide to passengers to help them sleep. These blindfolds both eliminate the subject's peripheral vision and make most subjects feel more relaxed by reducing distraction and creating a mild form of sensory deprivation.

The looker sits behind the subject, and in a series of twenty trials either looks at the back of the subject's neck, or looks away and thinks of something else. The sequence of trials is randomized. The simplest way to do this is for the looker to toss a coin before each trial: Heads mean "look," tails mean "don't look." Random number tables or a random number generator can be used instead, taking odd numbers to mean "look" and even numbers to mean "don't look"; or ready-made randomized instruction sheets can be downloaded from my Web site (www.sheldrake.org).

Just before each trial the looker signals to the subject by means of a mechanical click or beep that the trial is beginning. Within about 10 seconds the subject guesses "looking" or "not looking." These guesses are either right or wrong and are recorded on the score sheet. A sample score sheet is shown in figure A–1.

There are two ways in which this experiment can be done: with feedback or without. Either the looker tells the subject

STARING EXPERIMENT SCORE SHEET

Jason looking at *Sharon*

Date: *November 15, 2001*

Details: *Sharon blindfolded, signaling with clicker, distance apart: 3 yards*

Trial	Look/No	Result
1	Look	✔
2	No	✘
3	Look	✘
4	No	✔
5	Look	✘
6	Look	✔
7	Look	✔
8	Look	✘
9	No	✘
10	No	✔
11	Look	✔
12	No	✘
13	Look	✔
14	Look	✘
15	No	✔
16	Look	✔
17	No	✘
18	Look	✔
19	No	✔
20	No	✘

	Looking		Not Looking	
TOTALS	Right 7	Wrong 4	Right 4	Wrong 5

GRAND TOTALS	Right 11	Wrong 9

FIG. A–1

A sample score sheet from a staring experiment

immediately after each guess whether or not it is right, or the looker does not tell the subject whether or not the guess is correct. Both methods generally give significant positive results, but subjects tend to perform better when given feedback, if only because it makes doing the experiment more interesting.[12]

The total numbers of right and wrong guesses are added up separately for the looking and not-looking trials, as are the overall totals (see figure A–1). Then the results from a series of subjects are tabulated, as in figure A–2, which shows actual data from an experiment conducted at a school in London.

There are two ways to add up and analyze the data. The first is simply to add up the total numbers in each column. The second (suggested to me by Prof. Nicholas Humphrey) is to assign a score to each subject depending on whether there are more right than wrong guesses (+), more wrong than right guesses (–) or equal numbers of right and wrong guesses (=). The advantage of this method is that it gives an equal weight to every subject, whereas with the first method, a minority of subjects who score either very positively or very negatively can have a disproportionate influence on the totals. The total scores using both of these methods are shown in figure A–2.

For statistical analysis, the scores for right and wrong guesses in the looking (L), not-looking (NL), and totals columns can be compared using standard statistical tests, such as the chi-squared test. The null hypothesis is that the number of correct and incorrect guesses will be the same. In other words, if people guess at random, the average scores in all cases should be 50 percent correct and 50 percent incorrect. Using the +, –, and = method, the null hypothesis is that the number of + scores will equal the number of – scores. (For this method, the = scores can be ignored.)

Over and over again these tests have shown a characteristic pattern of results, whereby subjects score above chance levels in the looking trials and at levels close to chance in the not-looking trials.[13] For example the data in figure A–3, which combine tests carried out with both adults and schoolchildren, show a hugely significant positive effect in the looking trials with astronomical odds against chance (10^{25} to 1), while there is no significant difference in the not-looking trials. A similar pattern is apparent in the data in figure A–2.

A SAMPLE TALLY OF SCORES

Looker/subject	L right	L +/-/=	L wrong	NL right	NL +/-/=	NL wrong	Total right	Total +/-/=	Total wrong
Mimi/Emma	7	+	3	4	−	6	11	+	9
Emma/Mimi	12	+	2	4	+	2	16	+	4
Jayvanti/Rajvee	6	+	3	4	−	7	10	=	10
Rajvee/Jayvanti	6	=	6	5	+	3	11	+	9
Neela/Grace	10	+	2	4	=	4	14	+	6
Grace/Neela	9	+	5	2	−	4	11	+	9
Sam/Jessica	8	+	1	4	−	7	12	+	8
Jessica/Sam	5	=	5	6	+	4	11	+	9
Lucy/Alix	5	+	3	9	+	3	14	+	6
Alix/Lucy	3	−	5	5	−	7	8	−	12
Anna/Clare	6	+	4	3	−	7	9	−	11
Clare/Anna	7	+	4	4	−	5	11	+	9
Holly/Stella	7	+	3	7	+	3	14	+	6
Stella/Holly	5	=	5	4	−	6	9	−	11
Totals	96		51	65		68	161		119
Right %	65%			49%			58%		
Totals +/-/=	10+	1−	3=	5+	8−	1=	10+	3−	1=

FIG. A–2

A sample tally of scores from a staring experiment. Each looker/ subject pair carried out 20 trials. The numbers of correct and incorrect guesses are shown for both the looking (L) and not looking (NL) trials as well as for the totals. In addition, the subjects were scored + if they made more correct than incorrect guesses, − if they made more incorrect than correct guesses, and = if the number of correct and incorrect guesses were the same. This experiment was conducted at St. James' School for Senior Girls, London, with girls in Year Eleven (ages fifteen to sixteen) on November 20, 2001.

PERCENTAGES OF RIGHT AND WRONG GUESSES

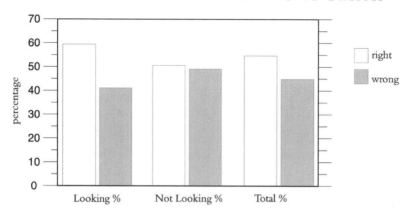

NUMBERS OF PEOPLE MORE RIGHT THAN WRONG OR MORE WRONG THAN RIGHT

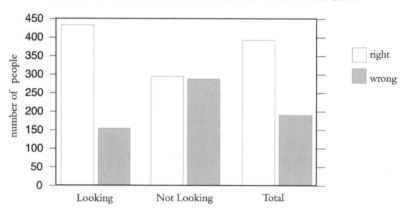

FIG. A–3

Combined results of staring experiments carried out with both adults and schoolchildren, using two different scoring systems. (Data from Sheldrake, 1999, Table 5.) The top graph depicts the percentages of right and wrong guesses in looking trials, not-looking trials, and totals. The bottom graph depicts the total numbers of people who were more right than wrong and more wrong than right in looking trials, not-looking trials, and totals. The statistical significance of the difference in the looking trials was 10^{25} to 1. It was non-significant in the not-looking trials and for the totals it was 10^{15} to 1.

This pattern is just what would be expected if the sense of being stared at is real. In looking trials most subjects have some ability to detect that they are being looked at. But in the not-looking trials, people are being asked to detect the *absence* of staring, for which there is no parallel in real life. We do not normally have a sense of *not* being stared at. Under these conditions, subjects are simply guessing.

This characteristic pattern of results serves as a built-in check against cheating, or against subjects picking up subtle clues from lookers, such as the sounds of head movements or changes in breathing. If people were cheating or picking up subtle sensory clues, they would be expected to score above chance levels in both the looking *and* the not-looking trials. But this is not what occurs.

Sensory clues were eliminated in experiments where blindfolded subjects were looked at through closed windows[14] or one-way mirrors,[15] and still the pattern of results was similar.

Subjects could also sense when they were being stared at in ordinary mirrors, when lookers could see only the subjects' reflected images.[16]

So far there have been few tests involving unusually sensitive subjects, but in a school in Germany some specially selected subjects had very high positive scores, up to 90 percent correct.[17]

There have so far been very few tests in which subjects have been tested repeatedly while being given feedback, to see how much they can improve through practice.

The largest experiment ever conducted on the sense of being stared at has been going on in Amsterdam, Holland, since 1995. Based on my suggestions for simple tests with looker-subject pairs, Diana Issidorides and her colleagues at the New Metropolis Center, a new science museum in Amsterdam, developed an ingenious computerized procedure that has made the experiment seem like a game with moving graphics and instructions. They have employed a sophisticated but user-friendly statistical methodology that gives immediate feedback on how the subject is doing.[18]

In this experiment, the looker sits behind the subject and is instructed whether or not to look by a signal on the computer screen.

For each trial, the subject guesses out loud and the looker enters the guess into the computer. Depending on the number of correct or incorrect guesses, after a maximum of thirty trials the computer announces whether or not the subject "has eyes in the back of the head."

The statistical program for this experiment is designed in such a way that if everyone were just guessing at random, 20 percent of the participants would be classified as having eyes in the back of their head. Against this chance expectation of 20 percent, in fact 32 to 40 percent of the subjects earned this designation. The percentages varied according to the subjects' age and sex. The most successful subjects were boys under the age of eight. More than 14,500 looker-subject pairs have taken part and the statistical significance of the positive results is astronomical: The odds against chance are 10^{462} to 1.

Since the first edition of this book was published, other investigators have continued to test for the sense of being stared at using closed-circuit television, as described on pages 117–18. They confirm the statistically significant positive results of the earlier reports.[19] The only exceptions to this general pattern are a few experiments carried out by skeptics in which the skeptical investigators themselves served as lookers.[20]

In my book *The Sense of Being Stared At, and Other Aspects of the Extended Mind,* due to be published in 2003, I summarize evidence for animals' sensitivity to being stared at. I discuss the evolution of this animal sensitivity and explore its implications for predator-prey relationships, and for the nature of perception and the extended mind.

More information about recent research on this subject, including the texts of my papers in scientific journals, is available on my Web site, www.sheldrake.org.

I welcome—either by e-mail through my Web site, or by post to one of the addresses given at the end of this appendix—the results of any experiments you participate in or carry out.

CHAPTER 5

THE REALITY OF PHANTOM LIMBS

With the help of Pam Smart, I have now developed a simpler and more effective procedure than the method described in chapter 5 to find out if people can detect the presence of a phantom limb.

Working with amputees with phantom arms, we carried out tests in which the amputee was on one side of a barrier, usually a closed door, while the person trying to detect the phantom (the "detector") was on the other. On both sides of the door, in corresponding positions, we fixed six sheets of paper, numbered 1 to 6, to mark out six zones. One experimenter (A) was on one side of the door with the amputee, while another experimenter (B) was on the other side with the detector.

For each trial, the experimenter with the amputee threw a die to select at random a number between 1 and 6. The amputee then "pushed" his or her phantom arm through the corresponding zone of the door. Experimenter A signaled that the trial was beginning by sounding a mechanical clicker. The phantom arm now extended through the door, and the detector felt with his or her hand in all six zones, trying to detect it. The detector then recorded on a score sheet the zone in which he or she believed the limb to be. Experimenter B signaled to A when the trial was completed, at which time Experimenter A threw the die again to begin a new trial. In a given session we usually conducted twenty trials. If the detectors had guessed at random, they would have been right about once out of six guesses.

Most of our experiments involved three or four detectors. They felt for the phantom arm one at a time, and noted their guesses on their score sheets in silence. The detectors did not talk to each other during the experiment and no feedback was given, so none knew which guesses were correct until all trials had been completed.

In a preliminary experiment, a single detector was right four times out of thirteen trials. By chance, he would have been right twice. In each of the four subsequent experiments there were twenty trials and three or four detectors. Out of the fourteen sets of data, one for each detector in each experiment, four were below the chance level and ten were above it. Altogether there was a total of 273 guesses. By chance, a sixth of these (16.7 percent) would have been correct. In fact, the success rate was 23.1 percent, significantly higher than the chance level ($p = 0.003$ by the binomial test). The most successful detector took part in three different experiments and had an overall success rate of 33.9 percent.

These results suggest that phantom limbs are indeed detectable. But the success rates were not very high overall. None of the people who volunteered to be detectors had any previous experience in trying to detect phantom limbs or in working with amputees. Perhaps detectors might be able to improve their sensitivity if tested repeatedly and given feedback after each guess.

The results of these initial experiments are encouraging and I hope that others will be able to do further research on this fascinating subject.

CHAPTER 6

THE VARIABILITY OF THE "FUNDAMENTAL CONSTANTS"

In the last few years, there have been unusually large variations in the values of the universal gravitational constant, G.[21]

As described in chapter 6, between 1970 and 1989 the "best" values of G varied between 6.6699 and 6.6745. More recently, very careful measurements at the German standards laboratory in Brunswick have given a value of 6.7154, an astonishing 0.6 percent higher than the previously accepted value.[22] Meanwhile measurements in other laboratories have given lower values: for example, at the University of Wuppertal, also in Germany, the figure was 6.6685.[23]

Usually laboratories publish only an average figure for G based on a series of measurements, after discarding values that seem to differ too much from the others. But in 1998, an American group working at the National Institute of Standards and Technology in

Boulder, Colorado published a series of data from measurements taken on different days, revealing a remarkable range of values. For example on one particular day the value of G was 6.73; on another day a few months later it was 6.64, 1.3 percent lower. [1]

Unfortunately, as far as I know, no one has yet tried to compare the measurements of G in different laboratories on the same days, as proposed in chapter 6, to see if these variations are correlated. If they are, it would suggest either that G is actually varying or that the measurements are affected by factors in the earth's environment that have hitherto been ignored. Either way, we would learn something new.

Meanwhile, there is now good evidence that at least one of the fundamental constants, the fine-structure constant, has changed as the universe has evolved. A careful series of measurements of light from very distant (and hence very old) quasars has shown that more than eight billion years ago the fine-structure constant was lower than it is today.[25] The change seems to have been small, about 1 part in 100,000, but nevertheless the theoretical consequences are enormous. The leader of the international team that made these measurements, John Webb of the University of New South Wales, Australia, cautiously commented, "It's possible that there is a time evolution of the laws of physics."[26]

Sheldon Glashow, a Nobel laureate in physics, remarked that if the results could be confirmed, the importance of such a discovery would rank "10 on a scale of 1 to 10."[27]

CHAPTER 7

THE EFFECTS OF EXPERIMENTERS' EXPECTATIONS

In most fields of science, researchers use blind methods rarely, if ever. The experimenters usually know which sample is which, and also know what kind of data to expect. By contrast, in psychology, parapsychology, and medicine it is generally recognized that blind procedures are necessary to guard against experimenters' expectations affecting the results they obtain. In experiments involving human subjects the trials are carried out double blind, so that neither the experimenter nor the human subject knows which treatment is which. For example, in a double blind clinical trial, neither the patients nor the clinicians know which patient has been given the drug being tested and which has been given a placebo.

I have carried out two quantitative surveys to measure how widely blind methodologies are used in different branches of science.

The first survey was of experimental papers published in recent issues of leading scientific journals, such as *Nature* and the *Proceedings of the National Academy of Sciences* (USA).[28] In the physical sciences, there were no blind experiments in any of the 237 papers reviewed. In the biological sciences, there were seven blind experiments out of 914 (0.8 percent); in psychology and animal behavior, seven out of 143 (4.9 percent); and in the medical sciences, fifty-five out of 227 (24.2 percent).[29] By far the highest proportion of blind experiments was in parapsychology, twenty-three out of twenty-seven (85.2 percent).

The second survey was of science departments at eleven British Universities (including Oxford, Cambridge, London, and Edinburgh). It confirmed that blind procedures were rare in most branches of the physical and biological sciences. They were neither used nor taught in twenty-two out of twenty-three physics and chemistry departments, or in fourteen out of sixteen biochemistry and molecular biology departments.[30] By contrast, blind methodologies were practiced and taught in four out of eight genetics departments, and in six out of eight physiology departments. Even so, in most of these departments they were used occasionally rather than routinely, and were mentioned only briefly in lectures.

Only in exceptional cases were blind techniques used routinely. This survey revealed three examples. All three involved commercial contracts, according to which the university scientists were required to analyze or evaluate coded samples without knowing their identity.

When academic scientists were interviewed for this survey, some did not know what was meant by the phrase "blind methodology."

Most of the others were aware of blind techniques but thought that they were necessary only in clinical research or psychology. They believed that their principal purpose was to avoid biases introduced by human subjects, rather than by experimenters. The most common view expressed by physical and biological scientists was that blind methodologies were unnecessary in the scientists' own fields because "nature itself is blind," as one professor put it. Some admitted the theoretical possibility of bias by experimenters but thought it of no importance in practice. One chemist added, "Science is difficult enough as it is without making it even harder by not knowing what you are working on."

The assumption by most "hard" scientists that blind techniques are unnecessary in their own field is so fundamental that it deserves to be put to the test, rather than taken for granted. Faith in the objectivity of science may turn out to be misplaced if researchers' expectations are regularly influencing what they find.

The only way to find out more is by means of empirical research. I suggest the following method,[31] which I illustrate with a typical experiment involving a test sample and a control sample. In biochemistry, for example, comparing an inhibited enzyme with an uninhibited control involves measuring the activity of both samples. Usually the experimenter knows which sample is which, and obviously expects the inhibited enzyme to be less active.

The test I propose is this: Carry out the experiment under the usual open conditions. But also do the same experiment under blind conditions, with the samples labeled A and B. In student practical classes, for instance, half the class would do the experiment blind, while the other half, in a separate laboratory, would know which sample is which, as usual.

If such tests show no disparity in the results under open and under blind conditions, then this would provide evidence that blind techniques make no difference. On the other hand, significant disparity in results under blind and open conditions would imply the existence of experimenter effects. Further research would then be needed to find out how the experimenters' expectations were influencing the data. Is the experimenter effect just a matter of selective reporting, or of selective observation and interpretation? Or could the experimenters' expectations influence what happens through a kind of psychokinesis or mind-over-matter effect?

The more independent investigations, the better. Perhaps it will turn out that "hard" scientists need not bother with blind techniques after all. On the other hand, perhaps professional scientists in the hard sciences are systematically biasing their results without knowing it.

CONTACT ADDRESSES

Please let me know of your findings in any of the areas of inquiry covered in this book. You may either e-mail me through my Web site, www.sheldrake.org., or you may get in touch by post at the following addresses.

In the United States:
Rupert Sheldrake
The Institute of Noetic Sciences
P.O. Box 6007
Petaluma, CA 94955-6007

In the United Kingdom:
Rupert Sheldrake
BM Experiments
London WC1N 3XX

A videotape based on this book, called *Seven Experiments That Could Change the World*, showing a number of the book's experiments actually being carried out, is now available commercially. The video is marketed in the United States by Wellspring Media Inc., and is available through Amazon.com.

NOTES

WHY BIG QUESTIONS DON'T NEED BIG SCIENCE

1. For an interesting discussion of this transition in Britain, see Berman (1974), "Hegemony and the amateur tradition in British science."
2. Kuhn (1970), *The Structure of Scientific Revolutions.*

WHY PUZZLING POWERS OF ANIMALS HAVE BEEN NEGLECTED

1. Popper and Eccles (1977), *The Self and Its Brain.*
2. For historical accounts of this controversy, see Sheldrake (1981), *A New Science of Life* and (1988) *The Presence of the Past.*
3. For example, the two leaders of the vitalist school in the early twentieth century, Hans Driesch and Henri Bergson, both served as presidents of the British Society for Psychical Research; and the vitalist views of the naturalist Eugene Marais enabled him to investigate the behavior of social animals in a very original way. His work on termites is discussed in chapter 3. And among psychical researchers there was a general openness to unusual powers in animals, as expressed, for example, by Haynes (1973) in *The Hidden Springs.*
4. Occam used this argument to oppose Platonists, with their notion of eternal, universal ideas existing either independently or as ideas in the mind of God. By the same token, this argument is opposed to the notion of universal mathematical laws of nature existing independently of human minds. Many mechanists, and certainly many physicists, are secret Platonists, and do not apply Occam's razor to this part of their thinking. Occam also used his razor against Aristotelians, with their doctrine of non-material essences inherent in material things. This argument would also rule out the real existence of fields, such as the universal gravitational and electromagnetic fields. Most mechanists do not take Occam's razor seriously in this case either, viewing the accepted fields of nature as really existing, rather than as mere models in the minds of physicists.
5. Some even see these questions in the light of a great struggle of good against evil, against "the Beast that slumbers below," as a

Harvard scientist, Gerald Holton, has put it. He recently exhorted the defenders of mechanistic science, which he describes as the "proper" worldview, to be on guard, calling upon them to try to "defang" this beast as "a duty they owe to their own belief system" (Holton, "How to think about the 'anti-science' phenomenon").

6. For a detailed discussion, see Sheldrake (1988), *The Presence of the Past*.

7. See, for example, Prigogine and Stengers (1984), *Order Out of Chaos*, Gleik (1988), *Chaos: Making a New Science,* and Waldrop (1993), *Complexity: The Emerging Science at the Edge of Order and Chaos*.

8. For a discussion of these developments and their implications, see Sheldrake (1990), *The Rebirth of Nature*.

CHAPTER 1
PETS WHO KNOW WHEN THEIR OWNERS ARE RETURNING

1. Long (1919), *How Animals Talk,* 78–9.
2. Ibid., 81–2.
3. Serpell (1986), *In the Company of Animals,* 103–4.
4. Ibid., 107.
5. *The New Penguin English Dictionary* (Harmondsworth: Penguin Books, 1986).
6. In the United States, the best-known such group is CSICOP, the Committee for the Scientific Investigation of Claims of the Paranormal. CSICOP stages annual conferences for skeptics and publishes a journal called *The Skeptical Inquirer*. Similar organizations have now been established in other countries and publish their own journals, such as *The British and Irish Skeptic*.
7. Serpell (1986), *In the Company of Animals,* 11–12.
8. Humphrey (1983), *Consciousness Regained*.
9. Woodhouse (1980), *Talking to Animals,* 202.
10. Smith (1989), *Animal Talk*.
11. Bardens (1987), *Psychic Animals*.
12. Ibid., 27.

CHAPTER 2
HOW DO PIGEONS HOME?

1. Genesis 8:8–11.
2. McFarland (1981), "Homing."
3. Inglis (1986), *The Hidden Power*.
4. Burnford (1961), *The Incredible Journey*.

5. Ibid.
6. Carthy (1963), *Animal Navigation*; Matthews, *Bird Navigation*.
7. Matthews (1968), *Bird Navigation*.
8. Carthy (1963), *Animal Navigation*.
9. Witherby 1938), *Handbook of British Birds*.
10. Baker (1980), *The Mystery of Migration*.
11. Berthold (1991), "Spatiotemporal programmes and the genetics of orientation."
12. Keeton (1981), "Orientation and Navigation of Birds."
13. Able (1982), "The effects of overcast skies on the orientation of free-flying nocturnal migrants."
14. Hasler (1978), Scholz and Horrall, "Olfactory imprinting and homing in salmon."
15. Gould (1990), "Why birds (still) fly south."
16. Schmidt-Koenig and Ganzhorn (1991), "On the problem of bird navigation."
17. Darwin (1859), *On the Origin of Species,* chapter 1; Darwin, *The Variation of Animals and Plants Under Domestication,* chapter 5.
18. Darwin (1873), "Origin of certain instincts."
19. Murphy (1873), "Instinct: A mechanical analogy."
20. Matthews (1968), *Bird Navigation*.
21. Ibid.
22. Wallraff (1990), "Navigation by homing pigeons."
23. Matthews (1968), *Bird Navigation*.
24. Keeton (1974), "The mystery of pigeon homing."
25. For example, Walraff (1990), "Navigation by homing pigeons."
26. Matthews (1968), *Bird Navigation,* 86.
27. Ibid., 87.
28. Osman and Osman (1976), *Pigeons in Two World Wars,* 83.
29. Schmidt-Koenig and Schlichte (1972), "Homing in pigeons with impaired vision."
30. Schmidt-Koenig (1979), *Avian Orientation and Navigation*.
31. Schmidt-Koenig and Schlichte (1972), "Homing in pigeons with Impaired Vision."
32. Schmidt-Koenig (1979), *Avian Orientation and Navigation,* 102.
33. Matthews (1968), *Bird Navigation*.
34. Keeton (1974), "The mystery of pigeon homing"; Lipp (1983), "Nocturnal homing in pigeons."
35. Schmidt-Koenig (1979), *Avian Orientation and Navigation*.
36. For details of the flight paths of clock-shifted birds, see Papi et al. (1991), "Homing strategies of pigeons."
37. Keeton (1981), "Orientation and navigation of birds."
38. Coemans and Vos (1992), "On the perception of polarized light by the homing pigeon."

39. An analysis of the long-range movement of aerosols also shows, in a more sophisticated way, the implausibility of this idea. However, in certain circumstances, the meteorological and atmospheric conditions may be more favorable for olfactory navigation over fairly short distances and in certain preferred directions, for example when there is a straight coastline and a regular pattern of sea breezes. Such conditions may prevail in Italy, where the main evidence favoring olfactory navigation has been collected. See Waldvogel (1987), "Olfactory navigation in homing pigeons."

40. Schmidt-Koenig (1987), "Bird navigation: Has olfactory orientation solved the problem?"

41. Matthews (1968), *Bird Navigation*.

42. Papi (1986), "Pigeon navigation: Solved problems and open questions" and "Olfactory navigation."

43. Papi et al. (1978), "Do American and Italian pigeons rely on different homing mechanisms?" For critical discussions of Papi's results, see Gould (1982), "The map sense of pigeons" and Schmidt-Koenig (1979), *Avian Orientation and Navigation*.

44. Keeton (1981), "Orientation and navigation of birds"; Gould (1982), "The map sense of pigeons"; Schmidt-Koenig (1979), *Avian Orientation and Navigation*.

45. For example, Papi (1982), "Olfaction and homing in pigeons."

46. Schmidt-Koenig (1979), *Avian Orientation and Navigation*; Wiltschko et al. (1987), "The orientation behaviour of anosmic pigeons in Frankfurt a.M., Germany." See also Wiltschko et al. (1987), "Pigeon homing."

47. Kiepenheuer et al. (1993), "Home-related and home-independent orientation of displaced pigeons."

48. Walcott (1991), "Magnetic maps in pigeons."

49. Matthews (1968), *Bird Navigation*.

50. For example, Wiltschko and Wiltschko (1976), "Die Bedeutung des Magnetikompasses für die Orientierung der Vögel" and (1991), "Orientation by the Earth's magnetic field in migrating birds and homing pigeons"; and Wiltschko (1993), "Magnetic compass orientation in birds and other animals."

51. Baker (1989), *Human Navigation and Magneto-Reception*.

52. Gould (1982), "The map sense of pigeons."

53. Schmidt-Koenig and Ganzhorn (1991), "On the problem of bird navigation"; for further examples, see Walcott (1989), "Show me the way you go home."

54. Schietecat (1990), "Pigeons and the weather"; Walcott (1991), "Magnetic maps in pigeons."

55. Wiltschko and Wiltschko (1988), "Magnetic orientation in birds";

Schmidt-Koenig and Ganzhorn (1991), "On the problem of bird navigation."

56. Walcott and Green (1974), "Orientation of homing pigeons altered by a change in the direction of an applied magnetic field."

57. Moore (1988), "Magnetic fields and orientation in homing pigeons."

58. Keeton (1972), "Effects of magnets on pigeon homing."

59. Moore (1988), "Magnetic fields and orientation in homing pigeons."

60. Moore, Stanhope and Wilcox (1987), "Pigeons fail to detect low-frequency magnetic fields." See also Papi et al. (1992), "Orientation-disturbing magnetic treatment affects the pigeon opioid system."

61. Walcott (1991), "Magnetic maps in pigeons, 49.

62. Schmidt-Koenig and Ganzhorn (1991), "On the problem of bird navigation."

63. Rhine (1951), "The present outlook on the question of psi in animals."

64. Pratt (1953), "The homing problem in pigeons" and (1956),"Testing for an ESP factor in pigeon homing."

65. Matthews (1968), *Bird Navigation*, 95–6.

66. Davies and Gribbin (1991), *The Matter Myth*, 217–18.

67. Thom (1975), *Structural Stability and Morphogenesis* and *Mathematical Models of Morphogenesis*; Abraham and Shaw (1984), *Dynamics: The Geometry of Behavior.*

68. Reprinted in Osman and Osman (1976), *Pigeons on Two World Wars.*

69. Ibid., 50.

70. Ibid.

71. Hill (1985), "Boomerang flying."

72. Hutton (1978), *Pigeon Lore.*

73. Personal communication, Dr Hans-Peter Lipp, of the University of Zürich-Irchel, the Officer Commanding Pigeons in the Swiss Army.

74. Spruyt (1950), *De Postduif van A–Z* (in Dutch) I am grateful to Louis van Gasteren for drawing my attention to this information and translating the relevant material.

75. Rhine and Feather (1962), "The study of cases of 'psi-trailing' in animals."

76. Ibid.

77. Rhine (1951), "The present outlook on the question of psi in animals," 241.

78. Rhine and Feather (1962), "The study of cases of 'psi-trailing' in animals," 17.

79. Long (1919), *How Animals Talk*, 95.

80. Ibid., 97–9.

CHAPTER 3

THE ORGANIZATION OF TERMITES

1. Baring and Cashford (1991), *The Myth of the Goddess*, 73.
2. von Frisch (1975), *Animal Architecture*, 123.
3. Griaule (1965), *Conversations with Ogotemmîli*, 17.
4. Evans-Pritchard 1937), *Witchcraft, Oracles and Magic Among the Azande*, 353.
5. Wilson (1971), *The Social Insects*.
6. Noirot 1970), "The nests of termites"; von Frisch *Animal Architecture*.
7. Wilson (1971), *The Social Insects*, 228.
8. Ibid., 317–19.
9. Ibid., 231.
10. For example, Wilson and Sober (1989), "Reviving the superorganism"; Seeley (1989), "The honey bee colony as superorganism"; Moritz and Southwick (1992), *Bees as Superorganisms*; Robinson (1993), "Colonial Rule."
11. An early proponent of this analogy was Hofstadter (1979), *Gödel, Escher, Bach*.
12. For example, Seeley and Levien (1987), "A colony of mind: The beehive as thinking machine"; Gordon et al. (1992), "A parallel distributed model of the behaviour of ant colonies."
13. Sole et al. (1993), "Oscillations and chaos in ant societies."
14. For a historical account of the morphogenetic field concept, see Sheldrake (1988), *The Presence of the Past*, chapter 6.
15. Sheldrake (1981), *A New Science of Life* and *The Presence of the Past*.
16. Stuart (1963), "Studies on the communication of alarm in the termite *Zootermopsis nevadensis*."
17. Stuart (1969), "Social behavior and communication."
18. H^lldobler and Wilson (1990), *The Ants*, 227.
19. Dunpert (1981), *The Social Biology of Ants*.
20. Stuart (1969), "Social behavior and communication"; Franks (1989), "Army ants: A collective intelligence"; H^lldobler and Wilson (1990), *The Ants*.
21. Wilson (1971), *The Social Insects*, 229.
22. Becker (1976), "Reaction of termites to weak alternating magnetic fields" and "Communication between termites by biofields."
23. Marais (1973), *The Soul of the White Ant*, 119–20.
24. Ibid., 121.
25. Ibid.,154.

INTRODUCTION TO PART TWO

CONTRACTED AND EXTENDED MINDS

1. See, for example, Palmer (1979), "A community mail survey of psychic experiences"; Haraldsson (1985), "Representative national surveys of psychic phenomenon"; Clarke (1991), "Belief in the paranormal"; Gallup and Newport (1991), "Belief in paranormal phenomenon among adult Americans."
2. Piaget (1973), *The Child's Conception of the World*, 70, 72, 78.
3. Whyte (1979), *The Unconscious Before Freud*.
4. Carus (1989), *Psyche: On the Development of the Soul*, 1.

CHAPTER 4

THE SENSE OF BEING STARED AT

1. Piaget (1973), *The Child's Conception of the World*, 61–2.
2. Ibid., chapter 1.
3. Poortman (1959), "The feeling of being stared at."
4. Conan Doyle, "J. Habakuk Jephson's Statement."
5. Haynes (1973), *The Hidden Springs*, 41.
6. London, *The Call of the Wild*, 77–8.
7. Long (1919), *How Animals Talk*, 91–2.
8. Following my visit in 1986 to the Mind Science Foundation, in San Antonio, Texas, we talked about the sense of being stared at and discussed various possible experiments. A project on the subject was initiated by William Braud and Sperry Andrews, the preliminary results of which are described in the text.
9. Elsworthy (1895), *The Evil Eye*.
10. Proverbs 28:22.
11. Heaton (1978), *The Eye: Phenomenology and Psychology of Function and Disorder*.
12. For an illuminating discussion of the mythological aspects of eyes and gazes, see Huxley (1990), *The Eye: The Seer and the Seen*.
13. Bacon, *Essays*, number 9.
14. Budge (1930), *Amulets and Superstitions*.
15. Titchener (1898), "The feeling of being stared at."
16. Ibid.
17. Coover (1913), "The feeling of being stared at."
18. Poortman (1959), "The feeling of being stared at."
19. Poortman did not analyze his data statistically, but I have analyzed them using the *t*-test (with the paired data sets) and the probability of the effect being due to chance is $p = 0.042$, below the $p = 0.05$

level conventionally used in assessment of significance.

20. Peterson (1978), "Through the looking glass: An investigation of the faculty of extra-sensory detection of being stared at."

21. Williams (1983), "The feeling of being stared at: A parapsychological investigation."

22. Braud et al. (1990), "Electrodermal correlates of remote attention: Autonomic reactions to an unseen gaze."; Braud, "Human interconnectedness: Research indications."

23. An analysis of the overall results from each of the ten experiments, using the paired t-test, shows a significance level of $p = 0.005$, indicating that there is only a 1 in 200 probability that the results were due to chance fluctuations.

24. The results can be analyzed statistically using the t-test, entering the total number of right and wrong guesses for each test as a data-pair.

25. The statistical significance was $p = 0.02$. See Mastrandrea (1991), "The feeling of being stared at").

26. For a discussion of some of them, see Abraham et al. (2001), *Chaos, Creativity, and Cosmic Consciousness,* chapter 5.

CHAPTER 5

THE REALITY OF PHANTOM LIMBS

1. Barja and Sherman (1985), *What to Expect When You Lose a Limb.*

2. James (1887), "The consciousness of lost limbs," 249.

3. Melzack (1992), "Phantom limbs."

4. Sherman et al. (1989), "The mystery of phantom pain."

5. Fischer (1969), "Out on a (phantom) limb"; Melzack (1989), "Phantom limbs, the self and the brain."

6. Melzack (1989), "Phantom limbs, the self and the brain."

7. Ibid.

8. Ibid.

9. Cited in Sacks (1985), *The Man Who Mistook His Wife for a Hat,* chapter 6.

10. Simmel (1956), "Phantoms in patients with leprosy and in elderly digital amputees."

11. Weinstein and Sarsen (1961), "Phantoms in cases of congenital absence of limbs"; Weinstein et al. (1964), "Phantoms and somatic sensation in cases of congenital aplasia."

12. Vetter and Weinstein (1967), "The history of the phantom in congenitally absent limbs."

13. Weinstein et al. (1964), "Phantoms and somatic sensation in cases of congenital aplasia."

14. Melzack (1992), "Phantom limbs."

15. Ibid.
16. Bromage and Melzack (1974), "Phantom limbs and the body schema."
17. Melzack and Bromage (1973), "Experimental phantom limbs"; Bromage and Melzack (1974), "Phantom limbs and the body schema."
18. Melzack and Bromage (1973), "Experimental phantom limbs," 263.
19. Ibid., 271.
20. Ibid.
21. Gross and Melzack (1978), "Body-image: Dissociation of real and perceived limbs by pressure-cuff ischemia."
22. Feldman (1940), "Phantom limbs."
23. Melzack (1992), "Phantom limbs," 120.
24. Mitchell (1872), *Injuries of Nerves and their Consequences,* 352.
25. Sacks (1985), *The Man Who Mistook His Wife for a Hat,* 66.
26. Barja and Sherman (1985), *What to Expect When You Lose a Limb.*
27. Exodus 21:24.
28. Mitchel (1872)l, *Injuries of Nerves and Their Consequences,* 357.
29. Frazer (1911), *The Golden Bough,* part 1, chapter 3, 52.
30. James (1887), "The consciousness of lost limbs."
31. Frazier and Kolb (1970), "Psychiatric aspects of the phantom limb."
32. Soloman and Schmidt (1978), "A burning issue: Phantom limb pain and psychological preparation of the patient for amputation."
33. For an admirably clear study of this phenomenon, see Green (1968), *Out-of-Body Experiences.*
34. Quoted in Blackmore (1983), *Beyond the Body: An Investigation of Out-of-Body Experiences,* 48.
35. Monroe (1973), *Journeys Out of the Body.*
36. Monroe (1985), *Far Journeys.*
37. Quoted in Moody (1976), *Life After Life,* 35.
38. Lorimer (1984), *Survival? Body, Mind and Death in the Light of Psychic Experience.* For collections of hundreds of cases, see Crookall (1961), *The Study and Practice of Astral Projection* ; (1964) *More Astral Projections* ; and (1972) *Case-Book of Astral Projection.*
39. Green (1968), *Lucid Dreams*; LaBerge (1985), *Lucid Dreaming.*
40. Melzack (1992), "Phantom limbs," 121.
41. Melzack (1989), "Phantom limbs, the self and the brain," 4.
42. For example, Karagalla and Kunz (1989), *The Chakras and the Human Energy Fields.*
43. Melzack (1992), "Phantom limbs."
44. Ibid.
45. Shreeve (1993), "Touching the phantom."
46. Ibid.

47. Melzack (1989), "Phantom limbs, the self and the brain," 9.

48. Ibid., 14.

49. Poeck and Orgass (1971), "The concept of the body schema."

50. Fischer (1969), "Out on a (phantom) limb."

51. Zuk (1956), "The phantom limb: A proposed theory of unconscious origins."

52. For technical details, see Dumitrescu (1983), *Electrographic Imaging in Medicine and Biology.*

53. Chaudhury et al. (1980), "Some advances in phantom leaf photography and identification of critical conditions for it."

54. Hubacher and Moss (1976), "The 'phantom leaf' effect as revealed through Kirlian photography"; Krippner (1980), *Human Possibilities: Mind Exploration in the USSR and Eastern Europe*; Stillings (1983), "The phantom leaf revisited: An interview with Allan Detrich."

55. Stanley Krippner, personal communication, 14 July 1993.

INTRODUCTION TO PART THREE

ILLUSIONS OF OBJECTIVITY

1. For an illuminating discussion, see Suzuki (1992), *Inventing the Future: Reflections on Science, Technology and Nature.*

2. Keller (1985), *Reflections on Gender and Science.*

3. Broad and Wade (1985), *Betrayers of the Truth, Fraud and Deceit in Science,* 197.

4. Gould (1984), *The Mismeasure of Man,* 27.

5. Medawar (1968), *The Art of the Soluble.*

6. Broad and Wade (1985), *Betrayers of the Truth: Fraud and Deceit in Science,* 27.

7. Westfall (1973), "Newton and the fudge factor."

8. Broad and Wade (1985), *Betrayers of the Truth: Fraud and Deceit in Science,* 34.

9. Ibid.

10. Ibid., 78.

11. Ibid., 141–2.

12. Ibid., 81.

13. Ibid.

14. Leviticus 16:20–2.

15. Broad and Wade (1985), *Betrayers of the Truth: Fraud and Deceit in Science,* 219.

16. Ibid., 218.

CHAPTER 6

THE VARIABILITY OF THE "FUNDAMENTAL CONSTANTS"

1. Petley (1985), *The Fundamental Physical Constants and the Frontiers of Metrology*.
2. Birge (1929), "Probable values of the general physical constants."
3. Discussed in Sheldrake (1988), *The Presence of the Past,* chapters 1 and 2.
4. See, for example, Wilber (1984), *Quantum Questions: Mystical Writings of the World's Great Physicists,* 101–11.
5. Pagels (1985), *Perfect Symmetry,* 11.
6. Barrow and Tipler (1986), *The Anthropic Cosmological Principle,* 5.
7. Davies (1992), *The Mind of God,* 221–2.
8. For example, Dirac (1974), "Cosmological models and the large numbers hypothesis."
9. Whitehead (1933), *Adventures of Ideas,* 143.
10. Sheldrake (1981), *A New Science of Life* ; (1988) *The Presence of the Past*; (1990) *The Rebirth of Nature*.
11. Gleik (1988), *Chaos: Making a New Science*.
12. Luther and Sagitor's measurements, respectively, cited in Petley (1985), *The Fundamental Physical Constants and the Frontiers of Metrology*.
13. Maddox (1986), "Turbulence assails fifth force."
14. Holding and Tuck (1984), "A new mine determination of the Newtonian gravitational constant."
15. For example, Holding et al. (1986), "Gravity in minesian investigation of Newton's law."
16. Fischbach et al. (1992), "Reanalysis of the Eötvös experiment."
17. Anderson (1988), "Icy tests provide firmer evidence for the fifth force"; Maddox (1988), "The stimulation of the fifth force."
18. Parker and Zumberge (1989), "An analysis of geophysical experiments to test Newton's law of gravity."
19. Fischbach and Talmadge (1992), "Six years of the fifth force."
20. Hellings, et al. (1983), "Experimental test of the variability of G using Viking Lander ranging data."
21. Reasenberg (1983), "The constancy of G and other gravitational experiments."
22. Damour, et al. (1988), "Limits on the variability of G using binary pulsar data."
23. For example, Wesson (1980), "Does gravity change with time?"; van Flandern (1981), "Is the gravitational constant changing?"
24. van Flandern (1981), "Is the gravitational constant changing?"
25. Petley (1985), *The Fundamental Physical Constants and the Frontiers of Metrology.* 46–50.

26. Ibid., 47–8.

27. "Light," *Encyclopaedia Britannica*, 15th ed.

28. Birge (1929), "Probable values of the general physical constants," 68.

29. de Bray (1934), "Velocity of Light."

30. Data from von Friesen (1937), "On the values of fundamental atomic constants," and Petley (1985), *The Fundamental Physical Constants and the Frontiers of Metrology*, 295.

31. Petley (1985), *The Fundamental Physical Constants and the Frontiers of Metrology*, 294–5.

32. Bearden and Thomsen (1951), "Résumé of atomic contants."

33. Petley (1985), *The Fundamental Physical Constants and the Frontiers of Metrology*, 68.

34. Barrow (1986), *The World Within the World*, 157.

35. von Friesen (1937), "On the values of fundamental atomic constants." 431.

36. Feynman (1985), *Surely You're Joking, Mr. Feynman*, 312–13.

37. von Friesen (1937), "On the values of fundamental atomic constants"; Birge, "A new table of the general physical constants."

38. Birge (1945), "The 1944 values of certain atomic constants with particular reference to the electronic charge."

39. Petley (1985), *The Fundamental Physical Constants and the Frontiers of Metrology*, 46; Barrow and Tipler (1986), *The Anthropic Cosmological Principle*, 241.

40. Cook (1957), "Secular changes of the units and constants of physics."

41. Birge (1929), "Probable values of the general physical constants" and (1941) "A new table of the general physical constants."

42. Barrow and Tipler (1986), *The Anthropic Cosmological Principle*.

43. Petley (1985), *The Fundamental Physical Constants and the Frontiers of Metrology*.

44. For example, Arp et al. (1990), "The extragalactic universe."

45. Gleik (1988), *Chaos: Making a New Science*.

46. Dousse and Rheme (1987), "A student experiment for the accurate measurement of the Newtonian gravitational constant."

47. For a comprehensive bibliography, see Gillies (1990), "Resource letter MNG-1: Measurements of Newtonian gravitation."

CHAPTER 7

THE EFFECTS OF EXPERIMENTERS' EXPECTATIONS

1. Lewis (1964), *The Discarded Image*.

2. Reber (1985), *The Penguin Dictionary of Psychology*, 317.

3. See, for example, Wolman (1977), *Handbook of Parapsychology.*

4. For example, the neurophysiologist Sir John Eccles favors this approach (Popper and Eccles (1977), *The Self and Its Brain*).

5. Rosenthal and Rubin (1978), "Interpersonal expectancy effects."

6. Rosenthal (1976), *Experimenter Effects in Behavior Research.*

7. For example, Rosenthal (1991), "Teacher expectancy effects."

8. Rosenthal and Rubin (1978), "Interpersonal expectancy effects," 412–13.

9. White, Tursky, and Schwartz (1985), *Placebo: Theory, Research and Mechanisms*; Murphy (1992), *The Future of the Body*, chapter 12.

10. Evans (1984), "Unravelling placebo effects."

11. From Psalm 114:9 in the Vulgate Bible, corresponding to Psalm 116:16 in the Authorized Version.

12. Shapiro (1970), "Placebo effect in psychotherapy and psychoanalysis."

13. Ibid.

14. Evans (1984), "Unravelling placebo effects," 17.

15. Quoted in Benson and McCallie (1979), "Angina pectoris and the placebo effect."

16. Evans (1984), "Unravelling placebo effects," 17.

17. Quoted in Dossey (1991), *Meaning and Medicine,* 203.

18. Benson and McCallie (1979), "Angina pectoris and the placebo effect."

19. Pogge (1963), "The toxic placebo."

20. S. Ross and L. W. Buckalew, in White, Tursky, and Schwartz (1985), *Placebo: Theory, Research and Mechanisms.*

21. Evans (1984), "Unravelling placebo effects."

22. Schweiger and Parducci (1978), "Placebo in Reverse."

23. R. A. Hahn, in White, Tursky and Schwartz (1985), *Placebo: Theory, Research and Mechanisms,* 182.

24. Rosenthal (1976), *Experimenter Effects in Behavioral Research,* chapter 10.

25. Ibid.

26. Ibid., 7.

27. Rosenthal (1976), *Experimenter Effects in Behavioral Research*; Rosenthal and Rubin (1978), "Interpersonal expectancy effects."

28. Quoted in Rosenthal (1976), *Experimenter Effects in Behavioral Research*, chapter 10.

29. Rhine (1934), *Extrasensory Perception.*

30. White (1976), "The influence of persons other than the experimenter on the subject's scores in psi experiments."

31. Kennedy and Taddonio (1976), "Experimenter effects in parapsychological research."

32. Schmidt (1973), "PK tests with a high-speed random number generator"; (1974)"Comparison of PK action on two different random number generators."

33. Honorton (1975), "Has science developed the confidence to confront claims of the paranormal?"

34. Honorton and Barksdale (1972), "PK performance with waking suggestions for muscle tension versus relaxation."

35. Stamford (1974), "An experimentally testable model for spontaneous psi occurrences."

36. Hasted et al. (1975), "Scientists confronting the paranormal."

37. Inglis (1986), *The Hidden Power,* 194.

38. Ibid., 195.

39. Rosenthal (1984), "Interpersonal expectancy effects and psi."

40. See, for example, Waddington (1957), "The genetic basis of the assimilated bithorax stock"; Ho et al. (1983), "Effects of successive generation of ether treatment."

GENERAL CONCLUSIONS

1. Popper (1959), *The Logic of Scientific Discovery,* 111.

APPENDIX TO THE SECOND EDITION

UPDATES ON THE SEVEN EXPERIMENTS

1. Sheldrake (1999), *Dogs That Know When Their Owners Are Coming Home;* Sheldrake and Smart (1998), "A dog that seems to know when its owner is returning"; (2000) "A dog that seems to know when its owner is coming home: Videotaped experiments and observations"; and (2000) "Testing a return-anticipating dog."

2. In the skeptics' tests, Jaytee was at the window for only 4 percent of the time when Pam was away from home, but for an average of 78 percent of the time when she was on her homewards journey. This difference was statistically significant (Sheldrake and Smart (2000), "A dog that seems to know when his owner is coming home: Videotaped experiments"). The skeptics themselves tried to disqualify Jaytee on the grounds that he visited the window briefly before Pam set off to come home (Wiseman, Smith and Milton, "Can animals detect when their owners are returning home?"). But they could only arrive at this negative conclusion by ignoring most of their own data (Sheldrake [1999], "Commentary on a paper by Wiseman, Smith, and Milton on the 'psychic pet' phenomenon"). See also Wiseman, Smith and Milton (2000), "The 'psychic pet' phenomenon: A reply to Rupert Sheldrake" and Sheldrake (2000), "The 'psychic pet' phenomenon."

3. Brown and Sheldrake (1998), "Perceptive pets: A survey in north-west California."
4. Sheldrake and Smart (1997), "Psychic pets."
5. Sheldrake, Lawlor, and Turney (1998), "Perceptive pets: A survey in London."
6. Sheldrake (1999), *Dogs That Know When Their Owners Are Coming Home,* Chapter 3.
7. The text of this discussion has been published in Kayzer (1997), *A Glorious Accident.*
8. A videocasette showing these experiments entitled *How Do Pigeons Home? The Mobile Loft Experiments* is available from Euro Television Productions, Kloveniersburgwal 49 1011 JX, Amsterdam.
9. The details of these experiments are described in a report by Nuboer (1996), "A test of Sheldrake's morphic field theory."
10. Van Oortmerssen, G. A., "Orientation of carrier pigeons at sea: A pilot study. Report from the Department of Animal Physiology, University of Groningen, Netherlands, 1997. (Original in Dutch).
11. Ibid.
12. Sheldrake (2001), "Experiments on the sense of being stared at."
13. Sheldrake (1998), "The sense of being stared at: Experiments in schools" and (1999) "The sense of being stared at confirmed by simple experiments."
14. Sheldrake (2000), "The sense of being stared at does not depend on known sensory clues."
15. See the data for experiment 1 in Colwell, Schröder, and Sladen (2000), "The ability to detect unseen staring." (For a discussion of their experiment 2, see Sheldrake (2000), "Research on the feeling of being stared at.")
16. Sheldrake (2003), forthcoming.
17. Sheldrake (1998), "The sense of being stared at: Experiments in schools."
18. The statistical procedure was developed by Jan van Bolhuis, a professor of statistics at the Amsterdam Free University.
19. Schlitz and LaBerge (1994), "Autonomic detection of remote observation" and (1997) "Covert observation increases skin conductance in subjects unaware of when they are being observed"; Wiseman and Schlitz (1997), "Experimenter effects and the remote detection of staring."
20. See Wiseman and Schlitz (1997), "Experimenter effects and the remote detection of staring." For references to other tests by skeptics, see the discussion in Sheldrake (2000), "Research on the feeling of being stared at."

21. Kiernan (1995), "Gravitational constant is up in the air"; Kestenbaum (1998), "Gravity measurement closes in on big G"; Watson (2000), "Getting a more precise grip on the physical world."

22. Quinn (2000), "Measuring big G."

23. Kiernan (1995), "Gravitational constant is up in the air."

24. Schwarz et al. (1998), "A free-fall determination of the Newtonian constant of gravity."

25. Collins (2001), "Plus ça change."

26. Glanz and Overbye (2001), "Cosmic laws like speed of light may be changing, a study finds."

27. Ibid.

28. Sheldrake (1998), "Experimenter effects in scientific research: How widely are they neglected?"

29. Sheldrake (1999), "How widely is blind assessment used in scientific research?" The percentage of blind papers in medical journals in this survey was higher than that previously reported in Sheldrake (1998), "Experimenter effects in scientific research: How widely are they neglected?" which involved only British medical journals. Including American journals in a larger sample raised the percentage, because there were more papers with blind methodologies in the American than in the British medical journals.

30. Sheldrake (1998), "Experimenter effects in scientific research: How widely are they neglected?"

31. Sheldrake (1998), "Could experimenter effects occur in the physical and biologial sciences?"

BIBLIOGRAPHY

Able, K. T. 1982. The effects of overcast skies on the orientation of free-flying nocturnal migrants. In *Avian Navigation,* edited by F. Papi and H. G. Wallraff. Berlin: Springer.

Abraham, R., and C. D. Shaw. 1984. *Dynamics: The Geometry of Behavior.* Santa Cruz: Aerial Press.

Abraham, R., T. McKenna, and R. Sheldrake. 2001. *Chaos, Creativity, and Cosmic Consciousness.* Rochester, Vt.: Park Street Press.

Anderson, I. August 29, 1988. Icy tests provide firmer evidence for the fifth force. *New Scientist* 11.

Arp, H. C., G. Burbidge, F. Hoyle, J. V. Narlikar, and N. C. Wickramasinghe. 1990. The extragalactic universe: An alternative view. *Nature* 346: 807–12.

Bacon, F. 1881. *Essays.* London: Macmillan.

Baker, R. R. 1980. *The Mystery of Migration.* London: Macdonald.

———. 1989. *Human Navigation and Magneto-Reception.* Manchester: Manchester University Press.

Bardens, D. 1987. *Psychic Animals: An Investigation of Their Secret Powers.* London: Hale.

Baring, A., and J. Cashford. 1991. *The Myth of the Goddess.* London: Viking.

Barja, R. H., and R. A. Sherman. 1985. *What to Expect When You Lose a Limb.* Fort Gordon, Ga.: Eisenhower Army Medical Center.

Barrow, J. D. 1988. *The World Within the World.* Oxford: Oxford University Press.

Barrow, J. D., and F. Tipler. 1986. *The Anthropic Cosmological Principle.* Oxford: Oxford University Press.

Bearden, J. A., and J. S. Thomsen. 1959. Résumé of atomic constants. *American Journal of Physics* 27: 569–76.

Bearden, J. A., and H. M. Watts. 1951. A re-evaluation of the fundamental atomic constants. *Physical Review* 21: 73–81.

Becker, G. 1976. Reaction of termites to weak alternating magnetic fields. *Naturwissenschaften* 63: 201.

———. 1977. Communication between termites by biofields. *Biological Cybernetics* 26: 41–51.

Benson, H., and D. McCallie. 1979. Angina pectoris and the placebo effect. *New England Journal of Medicine* 300: 1424–29.

Berman, M. 1974. Hegemony and the amateur tradition in British science. *Journal of Social History* 8: 30–50.

Berthold, P. 1991. Spatiotemporal programmes and the genetics of orientation. In *Orientation in Birds,* edited by P. Berthold. Basel: Birkhäuser.

Birge, R. T. 1929. Probable values of the general physical constants. *Reviews of Modern Physics* 1: 1–73.

———. 1941. A new table of the general physical constants. *Reviews of Modern Physics* 13: 233–39.

———. 1945. The 1944 values of certain atomic constants with particular reference to the electronic charge. *American Journal of Physics* 13: 63–73.

Blackmore, S. 1983. *Beyond the Body: An Investigation of Out-of-the-Body Experiences.* London: Paladin.

Braud, W. G. 1992. Human interconnectedness: Research indications. *ReVision* 14: 140–48.

Braud, W. G., D. Shafer, and S. Andrews. 1990. Electrodermal correlates of remote attention: Autonomic reactions to an unseen gaze. In *Proceedings of the Parapsychological Association 33rd Annual Convention, USA.* Metuchen, N.J.: Scarecrow Press.

Broad, W., and N. Wade. 1985. *Betrayers of the Truth: Fraud and Deceit in Science.* Oxford: Oxford University Press.

Bromage, P. R., and R. Melzack. 1974. Phantom limbs and the body schema. *Canadian Anaesthetists' Society Journal* 21: 267–74.

Brown, D. J., and R. Sheldrake, 1998. Perceptive pets: A survey in northwest California. *Journal of the Society for Psychical Research* 62: 396–406.

Budge, W. 1930. *Amulets and Superstitions.* Oxford: Oxford University Press.

Burnford, S. 1961. *The Incredible Journey.* London: Hodder and Stoughton.

Carthy, J. D. 1963. *Animal Navigation.* London: Unwin.

Carus, C. G. 1989. *Psyche: On the Development of the Soul.* Dallas: Spring Publications.

Chaudhury, J. K., P. C. Kejariwal, and A. Chattopadhyay. 1980. Some advances in phantom leaf photography and identification of critical conditions for it. Paper presented at the 4th Annual Conference of the International Kirlian Research Association, June 13–15. 1980.

Clarke, D. 1991. Belief in the paranormal: A New Zealand survey. *Journal of the Society for Psychical Research* 57: 412–25.

Coemans, M., and J. Vos. 1992. On the perception of polarized light by the homing pigeon. Doctoral thesis, University of Utrecht.

Cohen, E. R., J. W. M. DuMond, T. W. Layton, and J. S. Rollett. 1955. Analysis of variance of the 1952 data on the atomic constants and a new adjustment, 1955. *Reviews of Modern Physics* 27: 363–80.

Cohen, E. R., and B. N. Taylor. 1973. The 1973 least-squares adjustment of the fundamental constants. *Journal of Physical and Chemical Reference Data* 2: 663–734.

————. 1988. The 1986 CODATA recommended values of the fundamental physical constants. *Journal of Physical and Chemical Reference Data* 17: 1795–803.

Collins, G. P. November 2001. Plus ca change. *Scientific American* 285: 16–17.

Colwell, J, S. Schr^der, and D. Sladen. 2000. The ability to detect unseen staring: A literature review and empirical tests. *British Journal of Psychology* 91: 71–85.

Conan Doyle, A. 1956. J. Habakuk Jephson's Statement. In *The Conan Doyle Stories*. London: Murray.

Condon, E. U. 1967. Adjusted values of constants. In *Handbook of Physics,* 2d ed., edited by E. U. Condon and H. Odishaw. New York: McGraw Hill.

Cook, A. H. 1957. Secular changes of the units and constants of physics. *Nature* 160: 1194–95.

Coover, J. E. 1913. The feeling of being stared at. *American Journal of Psychology* 24: 570–75.

Crookall, R. 1961. *The Study and Practice of Astral Projection*. London: Aquarian Press.

————. 1964. *More Astral Projections*. London: Aquarian Press.

————. 1972. *Case-Book of Astral Projection*. Secaucus, N.J.: University Books.

Damour, T., G. W. Gibbons, and J. H. Taylor. 1988. Limits on the variability of G using binary pulsar data. *Physical Review Letters* 61: 1151–54.

Darwin, C. 1859. *On The Origin of Species by Means of Natural Selection*. London: Murray.

————. 1873. Origin of certain instincts. *Nature* 7: 417–18.

————. 1881. *The Variation of Animals and Plants Under Domestication*. London: Murray.

Davies, P. 1992. *The Mind of God*. London: Simon and Schuster.

Davies, P., and J. Gribbin. 1991. *The Matter Myth: Towards 21st-Century Science*. London: Viking.

de Bray, E. J. G. 1934. Velocity of light. *Nature* 133: 948.

Dirac, P. 1974. Cosmological models and the large numbers hypothesis. *Proceedings of the Royal Society* A338: 439–46.

Dossey, L. 1991. *Meaning and Medicine*. New York: Bantam.

Dousse, J. C., and C. Rheme. 1987. A student experiment for the accurate measurement of the Newtonian gravitational constant. *American Journal of Physics* 55: 706–11.

Dröscher, V. R. 1964. *Mysterious Senses*. London: Hodder and Stoughton.

Dumitrescu, I. F. 1983. *Electrographic Imaging in Medicine and Biology*. Suffolk: Neville Spearman.

Dunpert, K. 1981. *The Social Biology of Ants*. Boston: Pitman.

Elsworthy, F. 1895. *The Evil Eye*. London: Murray.

Evans, F. J. 1984. Unravelling placebo effects. *Advances: Institute for the Advancement of Health* 1, no. 3: 11–20.

Evans-Pritchard, E. E. 1937. *Witchcraft, Oracles and Magic Among the Azande*. Oxford: Oxford University Press.

Feldman, S. 1940. Phantom limbs. *American Journal of Psychology* 53: 590–92.

Feynman, R. 1985. *Surely You're Joking, Mr Feynman: Adventures of a Curious Character*. New York: Norton.

Fischbach, E., D. Sudarsky, A. Szafer, C. Talmadge, and S. H. Aronson. 1986. Reanalysis of the Eötvös experiment. *Physical Review Letters* 56: 3–6.

Fischbach, E., and C. Talmadge. 1992. Six years of the fifth force. *Nature* 356: 207–15.

Fischer, R. Winter 1969. Out on a (phantom) limb. *Perspectives in Biology and Medicine:* 259–73.

Franks, N. R. 1989. Army ants: A collective intelligence. *American Scientist* 77: 139–45.

Frazer, J. 1911. *The Golden Bough: Part I, The Magic Art and the Evolution of Kings*. London: Macmillan.

Frazier, S. H. and L. C. Kolb. 1970. Psychiatric aspects of the phantom limb. *Orthopedic Clinics of North America* 1: 481–95.

Gallup, G. H., and F. Newport. 1991. Belief in paranormal phenomena among adult Americans. *Skeptical Inquirer* 15: 137–46.

Gillies, G. T. 1990. Resource letter MNG-1: Measurements of Newtonian gravitation. *American Journal of Physics* 58: 525–34.

Glanz, J. and D. Overbye. August 16, 2001. Cosmic laws like speed of light may be changing, a study finds. *New York Times.*

Gleik, J. 1988. *Chaos: Making a New Science*. London: Heinemann.

Gordon, D. M., B. C. Goodwin, and L. E. H. Trainor. 1992. A parallel distributed model of the behaviour of ant colonies. *Journal of Theoretical Biology* 156: 293–307.

Gould, J. L. 1982. The map sense of pigeons. *Nature* 296: 205–11.

———. 1990. Why birds (still) fly south. *Nature* 347: 331.

Gould, S. J. 1984. *The Mismeasure of Man*. Harmondsworth: Pelican.

Green, C. 1968. *Lucid Dreams*. Oxford: Institute of Psychophysical Research.

———. 1968. *Out-of-the-Body Experiences*. Oxford: Institute of Psychophysical Research.

Griaule, M. 1965.
Conversations with Ogotemmîli. Oxford: Oxford University Press.

Gross, Y., and R. Melzack. 1978. Body-image: Dissociation of real and perceived limbs by pressure-cuff ischemia. *Experimental Neurology* 61: 680–88.

Haraldsson, E. 1985. Representative national surveys of psychic phenomena. *Journal of the Society for Psychical Research* 53: 145–58.

Hasler, A. D., A. T. Scholz, and R. M. Horrall. 1978. Olfactory imprinting and homing in salmon. *American Scientist* 66: 347–55.

Hasted, J. B., D. J. Bohm, E. W. Bastin, and B. O'Reagen. 1975. Scientists confronting the paranormal. *Nature* 254: 470–72.

Haynes, R. 1973. *The Hidden Springs: An Enquiry into Extra-Sensory Perception*. London: Hutchinson.

Heaton, J. M. 1978. *The Eye: Phenomenology and Psychology of Function and Disorder*. London: Tavistock Press.

Hellings, R. W., P. J. Adams, J. D. Anderson, M. S. Keesey, E. L. Lau, E. M. Standish, V. M. Canuto, and I. Goldman. 1983. Experimental test of the variability of G using Viking Lander ranging data. *Physical Review Letters* 51: 1609–12.

Hill, C. 1985. Boomerang flying. *Racing Pigeon Pictorial* 15: 116–18.

Hindley, J., and C. Rawson.1988. *How Your Body Works*. London: Usborne.

Ho, M. W., C. Tucker, D. Keeley, and P. T. Saunders. 1983. Effects of successive generations of ether treatment on penetrance and expression of the *Bithorax* phenocopy in *Drosophila melanogaster*. *Journal of Experimental Zoology* 225: 357–68.

Hofstadter, D. R. 1979. *Gödel, Escher, Bach: A Metaphorical Fugue of Minds and Machines*. Brighton: Harvester Press.

Holding, S. C., F. D. Stacey, and G. J. Tuck. 1986. Gravity in mines: An investigation of Newton's law. *Physical Review* D33. 3487–94.

Holding, S. C., and G. J. Tuck. 1984. A new mine determination of the Newtonian gravitational constant. *Nature* 307: 714–16.

Hölldobler, B., and E. O. Wilson. 1990. *The Ants*. Berlin: Springer.

Holton, G. 1992. How to think about the "anti-science" phenomenon. *Public Understanding of Science* 1: 103–28.

Honorton, C. 1975. Has science developed the confidence to confront claims of the paranormal? In *Research in Parapsycholopgy*, edited by J. D. Morris et al. Metuchen, N.J.: Scarecrow Press.

Honorton, C., and W. Barksdale. 1972. PK performance with waking suggestions for muscle tension versus relaxation. *Journal of the American Society for Psychical Research* 66: 208–12.

Hubacher, J., and T. Moss. 1976. The "phantom leaf" effect as revealed through Kirlian photography. *Psychoenergetic Systems* 1: 223–32.

Humphrey, N. 1983. *Consciousness Regained: Chapters in the Development of Mind*. Oxford: Oxford University Press.

Hutton, A. N. 1978. *Pigeon Lore*. London: Faber and Faber.

Huxley, F. 1990. *The Eye: The Seer and the Seen*. London: Thames and Hudson.

Inglis, B. 1986. *The Hidden Power.* London: Jonathan Cape.

James, W. 1887. The consciousness of lost limbs' *Proceedings of the American Society for Psychical Research* 1: 249–58.

Kahn, F. 1949. *The Secret of Life: The Human Machine and How It Works.* London: Odhams.

Karagalla, S., and D. Kunz. 1989. *The Chakras and the Human Energy Fields.* Wheaton, Ill: Quest Books.

Kayzer, W., ed. 1997. *A Glorious Accident.* New York: W. H. Freeman.

Keeton, W. T. 1972. Effects of magnets on pigeon homing. In *Animal Orientation and Navigation,* edited by S. R. Galler et al. Washington, DC: NASA.

———. December, 1974. The mystery of pigeon homing. *Scientific American.*

———. 1981. Orientation and navigation of birds. In *Animal Migration,* edited by D. J. Aidley. Society for Experimental Biology Seminar Series 13. Cambridge: Cambridge University Press.

Keller, E. F. 1985. *Reflections on Gender and Science.* New Haven: Yale University Press.

Kennedy, J. E., and J. L. Taddonio. 1976. Experimenter effects in parapsychological research. *Journal of Parapsychology* 40: 1–33.

Kestenbaum, D. 1998. Gravity measurement closes in on big G. *Science* 282: 2180–81.

Kiepenheuer, J., M. F. Neumann, and H. G. Wallraff. 1993. Home-related and home-independent orientation of displaced pigeons with and without olfactory access to environmental air. *Animal Behaviour* 45: 169–82.

Kiernan, V. April 26, 1995. Gravitational constant is up in the air. *New Scientist* 18.

Krippner, S. 1980. *Human Possibilities: Mind Exploration in the USSR and Eastern Europe.* New York: Doubleday.

Kuhn, T. S. 1970. *The Structure of Scientific Revolutions,* 2nd ed. Chicago: University of Chicago Press.

LaBerge, S. 1985. *Lucid Dreaming.* Los Angeles, Calif.: Tarcher.

Lewis, C. S. 1964. *The Discarded Image.* Cambridge: Cambridge University Press.

Lipp, H. P. 1983. Nocturnal homing in pigeons. *Comparative Biochemistry and Physiology* 76A: 743–49.

London, J. 1991. *The Call of the Wild.* London: Mammoth.

Long, W. J. 1919. *How Animals Talk.* New York: Harper.

Lorimer, D. 1984. *Survival? Body, Mind and Death in the Light of Psychic Experience.* London: Routledge and Kegan Paul.

McFarland, D. 1981. Homing. In *The Oxford Companion to Animal Behaviour,* edited by D. McFarland. Oxford: Oxford University Press.

Maddox, J. 1986. Turbulence assails fifth force. *Nature* 323: 665.

————. 1988. The stimulation of the fifth force. *Nature* 335: 393.

Marais, E. 1973. *The Soul of the White Ant*. Harmondsworth: Penguin.

Marks, D. and J. Colwell. September/October, 2000. The psychic staring effect: An artifact of pseudo-randomization. *Skeptical Inquirer:* 41–49.

Mastrandrea, M. 1991. The feeling of being stared at. Project report, Neuva Middle School, Hillsborough, Calif.

Matthews, G. V. T. 1968. *Bird Navigation,* 2d ed. Cambridge: Cambridge University Press.

Medawar, P. 1968. *The Art of the Soluble*. London: Methuen.

Melzack, R. 1989. Phantom limbs, the self and the brain. *Canadian Psychology* 30: 1–16.

————. April, 1992. Phantom limbs. *Scientific American:* 120–26.

Melzack, R., and P. R. Bromage. 1973. Experimental phantom limbs. *Experimental Neurology* 39: 261–69.

Mitchell, S. W. 1872. *Injuries of Nerves and their Consequences*. Philadelphia: Lippincott.

Monroe, R. A. 1973. *Journeys Out of the Body*. New York: Doubleday.

————. 1985. *Far Journeys*. New York: Doubleday.

Moody, R. A. 1976. *Life After Life*. New York: Bantam.

Moore, B. R. 1988. Magnetic fields and orientation in homing pigeons: Experiments of the late W. T. Keeton. *Proceedings of the National Academy of Sciences, USA* 85: 4907–09.

Moore, B. R., K. J. Stanhope, and D. Wilcox. 1987. Pigeons fail to detect low-frequency magnetic fields. *Animal Learning and Behavior* 15: 115–17.

Moritz, R. F. A., and E. F. Southwick. 1992. *Bees as Superorganisms: An Evolutionary Reality*. Berlin: Springer.

Murphy, J. J. 1873. Instinct: A mechanical analogy. *Nature* 7: 483.

Murphy, M. 1992. *The Future of the Body*. Los Angeles, Calif.: Tarcher.

Noirot, C. 1970. The nests of termites. In *The Biology of Termites,* vol 2, edited by K. Krishna and F. M. Weesner. New York: Academic Press.

Nuboer, W. 1996. A test of Sheldrake's morphic field theory. Internal Publication of the Helmholtz Institute, University of Utrecht, Holland.

Osman, A. H., and W. H. Osman. 1976. *Pigeons in Two World Wars*. London: Racing Pigeon Publishing Co.

Pagels, H. 1985. *Perfect Symmetry*. London: Michael Joseph.

Palmer, J. 1979. A community mail survey of psychic experiences. *Journal of the American Society for Psychical Research* 73: 221–51.

Papi, F. 1982. Olfaction and homing in pigeons: Ten years of experiments. In *Avian Navigation,* edited by F. Papi and H. G. Wallraff. Berlin: Springer

————. 1986. Pigeon navigation: Solved problems and open questions. *Monitore Zoologia Italiana (NS)* 20: 471–517.

————. 1991. Olfactory navigation. In *Orientation in Birds,* edited by P. Berthold. Basel: Birkh%oouser.

Papi, F., P. Ioale, P. Dall'Antonia, and S. Benvenuti. 1991. Homing strategies of pigeons investigated by clock shift and flight path reconstruction. *Naturwissenschaften* 78: 370–73.

Papi, F., W. T. Keeton, A. I. Brown, and S. Benvenuti. 1978. Do American and Italian pigeons rely on different homing mechanisms? *Journal of Comparative Physiology* 128: 303–17.

Papi, F., P. Luschi, and P. Limonta. 1992. Orientation-disturbing magnetic treatment affects the pigeon opioid system. *Journal of Experimental Biology* 166: 169–79.

Parker, R. L., and M. A. Zumberge. 1989. An analysis of geophysical experiments to test Newton's law of gravity. *Nature* 342: 29–32.

Peterson, D. 1978. Through the looking glass: An investigation of the faculty of extra-sensory detection of being stared at. Masters thesis, Department of Psychology, University of Edinburgh.

Petley, B. W. 1985. *The Fundamental Physical Constants and the Frontiers of Metrology.* Bristol: Adam Hilger.

Piaget, J. 1973. *The Child's Conception of the World.* London: Granada.

Poeck, K., and B. Orgass. 1971. The concept of the body schema: A critical review and some experimental results. *Cortex* 7: 254–77.

Pogge, R. C. 1963. The toxic placebo. *Medical Times* 91: 773–81.

Poortman, J. J. 1959. The feeling of being stared at. *Journal of the Society for Psychical Research* 40: 4–12.

Popper, K. 1959. *The Logic of Scientific Discovery.* London: Hutchinson.

Popper, K., and J. Eccles. 1977. *The Self and its Brain.* Berlin: Springer.

Pratt, J. G. 1953. The homing problem in pigeons. *Journal of Parapsychology* 17: 34–60.

————. 1956. Testing for an ESP factor in pigeon homing. *Ciba Foundation Symposium on Extrasensory Perception.* Ciba Foundation, London.

Prigogine, I., and I. Stengers. 1984. *Order Out of Chaos.* Heinemann: London.

Quinn, T. 2000. Measuring big G. *Nature* 408: 919-20.

Reasenberg, R. D. 1983. The constancy of G and other gravitational experiments. *Philosophical Transactions of the Royal Society* A310: 227–38.

Reber, A. S. 1985. *The Penguin Dictionary of Psychology.* Harmondsworth: Penguin.

Rhine, J. B. 1934. *Extrasensory Perception.* Boston: Boston Society for Psychical Research.

————. 1951. The present outlook on the question of psi in animals. *Journal of Parapsychology* 15: 230–51.

Rhine, J. B., and S. R. Feather. 1962. The study of cases of "psitrailing" in animals. *Journal of Parapsychology* 26: 1–22.

Robinson, G. E. 1993. Colonial rule. *Nature* 362: 126.

Rosenthal, R. 1976. *Experimenter Effects in Behavioral Research*. New York: John Wiley.

————. 1984. Interpersonal expectancy effects and psi: Some commonalties and differences. *New Ideas in Psychology* 2: 47–50.

————. 1991. Teacher expectancy effects: A brief update 25 years after the Pygmalion experiment. *Journal of Research in Education* 1: 3–12.

Rosenthal, R. and D. B. Rubin. 1978. Interpersonal expectancy effects: The first 345 studies. *Behavioral and Brain Sciences* 3: 377–415.

Sacks, O. 1985. *The Man Who Mistook his Wife for a Hat*. London: Duckworth.

Schietecat, G. 1990. Pigeons and the weather. *The Natural Winning Ways* 10: 13–22.

Schlitz, M. and S. LaBerge. 1994. Autonomic detection of remote observation: Two conceptual replications. *Proceedings of Presented Papers, Parapsychology Association 37th Annual Convention*, Amsterdam: 352–60.

————. 1997. Covert observation increases skin conductance in subjects unaware of when they are being observed: A replication. *Journal of Parapsychology* 61: 185–195.

Schmidt, H. S. 1973. PK tests with a high-speed random number generator. *Journal of Parapsychology* 37: 115–18.

————. 1974. Comparison of PK action on two different random number generators. *Journal of Parapsychology* 38: 47–55.

Schmidt-Koenig, K. 1979. *Avian Orientation and Navigation*. London: Academic Press.

————. 1987. Bird navigation: Has olfactory orientation solved the problem? *Quarterly Review of Biology* 62: 33–47.

Schmidt-Koenig, K., and J. U. Ganzhorn. 1991. On the problem of bird navigation. In *Perspectives in Ethology,* vol. 9, edited by P. P. G. Bateson and P. H. New York: Klopfer.

Schmidt-Koenig, K., and H. J. Schlichte. 1972. Homing in pigeons with impaired vision. *Proceedings of the National Academy of Sciences, USA* 69: 2446–47.

Schwarz, J. P., D. S. Robertson, T. M. Niebauer, and J. E. Faller. 1998. A free-fall determination of the Newtonian constant of gravity. *Science* 282: 2230–34.

Schweiger, A., and A. Parducci. 1978. Placebo in reverse. *Brain/Mind Bulletin* 3, no. 23: 1.

Seeley, T. D. 1989. The honey bee colony as superorganism. *American Scientist* 77: 546–53.

Seeley, T. D., and R. A. Levien. 1987. A colony of mind: The beehive as thinking machine. *The Sciences* 27: 38–43.

Serpell, J. 1986. *In the Company of Animals*. Oxford: Basil Blackwell.

Shapiro, A. K. 1970. Placebo effect in psychotherapy and psychoanalysis. *Journal of Clinical Pharmacology* 10: 73–77.

Sheldrake, R. 1981. *A New Science of Life: The Hypothesis of Formative Causation.* London: Blond and Briggs.

———. 1988. *The Presence of the Past: Morphic Resonance and the Habits of Nature.* London: Collins.

———. 1990. *The Rebirth of Nature: The Greening of Science and God.* London: Century.

———. May/June 1998. Could experimenter effects occur in the physical and biological sciences? *Skeptical Inquirer:* 57–8.

———. 1998. Experimenter effects in scientific research: How widely are they neglected? *Journal of Scientific Exploration* 12: 73–78.

———. 1998. The sense of being stared at: Experiments in schools. *Journal of the Society for Psychical Research* 62: 311–23.

———. 1999. Commentary on a paper by Wiseman, Smith, and Milton on the "psychic pet" phenomenon. *Journal of the Society for Psychical Research* 63: 306–11.

———. 1999. *Dogs That Know When Their Owners Are Coming Home, and Other Unexplained Powers of Animals.* London: Hutchinson.

———. 1999. How widely is blind assessment used in scientific research? *Alternative Therapies* 5: 88–91.

———. 1999. The sense of being stared at' confirmed by simple experiments. *Biology Forum* 92: 53–76.

———. 2000. The "psychic pet" phenomenon. *Journal of the Society for Psychical Research* 64: 126–28.

———. March/April 2000. Research on the feeling of being stared at. *Skeptical Inquirer:* 58–61.

———. 2000. The sense of being stared at does not depend on known sensory clues. *Biology Forum* 93: 209–24.

———. 2001. Experiments on the sense of being stared at: The elimination of possible artifacts. *Journal of the Society for Psychical Research* 65: 122–37.

Sheldrake, R., C. Lawlor, and J. Turney. 1998. Perceptive pets: A survey in London. *Biology Forum* 91: 57–74.

Sheldrake, R., and P. Smart. 1997. Psychic pets: A survey in northwest England. *Journal of the Society for Psychical Research* 61: 353–64.

———. 1998. A dog that seems to know when its owner is returning: Preliminary investigations. *Journal of the Society for Psychical Research* 62: 220–32.

———. 2000. A dog that seems to know when his owner is coming home: Videotaped experiments and observations. *Journal of Scientific Exploration* 14: 233–55.

———. 2000. Testing a return-anticipating dog, Kane. *Anthrozoos* 13: 203–12.

Sherman, R. A., J. C. Arena, C. J. Sherman, and J. L. Ernst. 1989. The mystery of phantom pain: Growing evidence for psychophysiological mechanisms. *Biofeedback and Self-Regulation* 14: 267–80.

Shreeve, J. June 1993. Touching the phantom. *Discover:* 35–42.

Simmel, M. L. 1956. Phantoms in patients with leprosy and in elderly digital amputees. *American Journal of Psychology* 69: 529–45.

Skaite, S. H. 1961. *The Study of Ants.* London: Longman

Smith, P. 1989. *Animal Talk: Interspecies Telepathic Communication.* Point Reyes Station, Calif.: Pegasus Publications.

Sole, R. V., O. Miramontes, and B. C. Goodwin. 1993. Oscillations and chaos in ant societies. *Journal of Theoretical Biology* 161: 343–57.

Soloman, G. F., and K. M. Schmidt. 1978. A burning issue: Phantom limb pain and psychological preparation of the patient for amputation. *Archives of Surgery* 113: 185–6.

Spruyt, C. A. M. 1950. *De Postduif van A–Z.* Gravenhage: Van Stockum.

Stamford, R. G. 1974. An experimentally testable model for spontaneous psi occurrences. *Journal of the American Society for Psychical Research* 66: 321–56.

Stillings, D. 1983. The phantom leaf revisited: An interview with Allan Detrich. *Archaeus* 1: 41–51.

Stuart, A. M. 1963. Studies on the communication of alarm in the termite *Zootermopsis nevadensis. Physiological Zoology* 36: 85–96.

———. 1969. Social behavior and communication. In *The Biology of Termites,* vol. 1, edited by K. Krishnaand and F. M. Weesner. New York: Academic Press.

Suzuki, D. 1992. *Inventing the Future: Reflections on Science, Technology and Nature.* London: Adamantine Press.

Thom, R. 1975. *Structural Stability and Morphogenesis.* Reading, Mass.: Benjamin.

———. 1983. *Mathematical Models of Morphogenesis.* Chichester: Horwood.

Titchener, E. B. 1898. The "feeling of being stared at." *Science New Series* 8: 895–97.

van Flandern, T. C. 1981. Is the gravitational constant changing? *Astrophysical Journal* 248: 813–18.

Vetter, R. J., and S. Weinstein. 1967. The history of the phantom in congenitally absent limbs. *Neuropsychologia* 5: 335–38.

von Friesen, S. 1937. On the values of fundamental atomic constants. *Proceedings of the Royal Society* A160: 424–40.

von Frisch, K. 1975. *Animal Architecture.* London: Hutchinson.

Waddington, C. H. 1957. The genetic basis of the assimilated bithorax stock. *Journal of Genetics* 55: 241–5.

Walcott, C. 1989. Show me the way you go home. *Natural History* 11: 40–6.

———. 1991. Magnetic maps in pigeons. In *Orientation in Birds,* edited by P. Berthold. Basel: Birkhäuser

Walcott, C., and R. P. Green. 1974. Orientation of homing pigeons altered by a change in the direction of an applied magnetic field. *Science* 184: 180–82.

Waldrop, M. M. 1993. *Complexity: The Emerging Science at the Edge of Order and Chaos.* London: Viking.

Waldvogel, J. A. 1987. Olfactory navigation in homing pigeons: Are the current models atmospherically realistic? *The Auk* 104: 369–79.

Wallraff, H. G. 1990. Navigation by homing pigeons. *Ethology, Ecology and Evolution* 2: 81–115.

Watson, A. 2000. Getting a more precise grip on the physical world. *Science* 287: 1391–93.

Weinstein, S., and E. A. Sarsen. 1961. Phantoms in cases of congenital absence of limbs. *Neurology* 11: 905–11.

Weinstein, S., E. A. Sarsen, and R. J. Vetter. 1964. Phantoms and somatic sensation in cases of congenital aplasia. *Cortex* 1: 276–90.

Wesson, P. S. 1980. Does gravity change with time? *Physics Today* 33: 32–37.

Westfall, R. S. 1973. Newton and the fudge factor. *Science* 180: 1118.

White, I., B. Tursky, and G. Schwartz, ed. 1985. *Placebo: Theory, Research and Mechanisms.* New York: Guilford Press.

White, R. 1976. The influence of persons other than the experimenter on the subject's scores in psi experiments. *Journal of the American Society for Psychical Research* 70: 132–66.

Whitehead, A. N. 1933. *Adventures of Ideas.* Cambridge: Cambridge University Press.

Whyte, L. L. 1979. *The Unconscious Before Freud.* London: Friedmann.

Wilber, K., ed. 1984. *Quantum Questions: Mystical Writings of the World's Great Physicists.* Boulder, Colo.: Shambala.

Williams, L. 1983. The feeling of being stared at: A parapsychological investigation. Bachelors thesis, Department of Psychology, University of Adelaide, South Australia. An abstract of this work was published in *Journal of Parapsychology* 47 (1983): 59.

Wilson, D. S., and E. Sober. 1989. Reviving the superorganism. *Journal of Theoretical Biology* 136: 337–56.

Wilson, E. O. 1971. *The Social Insects.* Cambridge, Mass.: Harvard University Press.

Wiltschko, W. 1993. Magnetic compass orientation in birds and other animals. In *Orientation and Navigation: Birds, Humans and Other Animals.* London: Royal Institution of Navigation.

Wiltschko, W., and R. Wiltschko. 1976. Die Bedeutung des Magnetikompasses für die Orientierung der Vögel. *Journal of Ornithology* 117: 363–87.

————. 1988. Magnetic orientation in birds. In *Current Ornithology*, vol. 5, edited by R. F. Johnston. New York: Plenum Press.

————. 1991. Orientation by the Earth's magnetic field in migrating birds and homing pigeons. *Progress in Biometeorology* 8: 31–43.

Wiltschko, W., R. Wiltschko, and M. Jahnel. 1987. The orientation behaviour of anosmic pigeons in Frankfurt a. M., Germany. *Animal Behaviour* 35: 1328–33.

Wiltschko, W., R. Wiltschko, and C. Walcott. 1987. Pigeon homing: Different aspects of olfactory deprivation in different countries. *Behavioral Ecology and Sociobiology* 21: 333–42.

Wiseman, R., and M. Schlitz. 1997. Experimenter effects and the remote detection of staring. *Journal of Parapsychology* 61: 197–207.

Wiseman, R., M. Smith, and J. Milton. 1998. Can animals detect when their owners are returning home? An experimental test of the "psychic pet" phenomenon. *British Journal of Psychology* 89: 453–462.

Wiseman, R., M. Smith, and J. Milton. 2000. The "psychic pet" phenomenon: A reply to Rupert Sheldrake. *Journal of the Society for Psychical Research* 64: 46–49.

Witherby, H. F. 1938. *Handbook of British Birds,* vol. 2. London: Witherby.

Wolman, B. B., ed. 1977. *Handbook of Parapsychology.* New York: Van Nostrand Reinhold.

Woodhouse, B. 1980. *Talking to Animals.* London: Allen Lane.

Zuk, G. H. 1956. The phantom limb: A proposed theory of unconscious origins. *Journal of Nervous and Mental Disorders* 124: 510–13.

INDEX

BOOKS OF RELATED INTEREST

THE PRESENCE OF THE PAST
Morphic Resonance and the Habits of Nature
by Rupert Sheldrake

A NEW SCIENCE OF LIFE
The Hypothesis of Morphic Resonance
by Rupert Sheldrake

THE REBIRTH OF NATURE
The Greening of Science and God
by Rupert Sheldrake

CHAOS, CREATIVITY, AND COSMIC CONSCIOUSNESS
by Rupert Sheldrake, Terence McKenna, and Ralph Abraham

THE BIOLOGY OF TRANSCENDENCE
A Blueprint of the Human Spirit
by Joseph Chilton Pearce

ALTERNATIVE SCIENCE
Challenging the Myths of the Scientific Establishment
by Richard Milton

ORIGINAL WISDOM
Stories of an Ancient Way of Knowing
by Robert Wolff

THE DREAMS OF DRAGONS
An Exploration and Celebration of the Mysteries of Nature
by Lyall Watson

Inner Traditions • Bear & Company
P.O. Box 388
Rochester, VT 05767
1-800-246-8648
www.InnerTraditions.com

Or contact your local bookseller